SpringerBriefs in Applied Sciences and Technology

SpringerBriefs present concise summaries of cutting-edge research and practical applications across a wide spectrum of fields. Featuring compact volumes of 50 to 125 pages, the series covers a range of content from professional to academic.

Typical publications can be:

- A timely report of state-of-the art methods
- An introduction to or a manual for the application of mathematical or computer techniques
- A bridge between new research results, as published in journal articles
- A snapshot of a hot or emerging topic
- An in-depth case study
- A presentation of core concepts that students must understand in order to make independent contributions

SpringerBriefs are characterized by fast, global electronic dissemination, standard publishing contracts, standardized manuscript preparation and formatting guidelines, and expedited production schedules.

On the one hand, **SpringerBriefs in Applied Sciences and Technology** are devoted to the publication of fundamentals and applications within the different classical engineering disciplines as well as in interdisciplinary fields that recently emerged between these areas. On the other hand, as the boundary separating fundamental research and applied technology is more and more dissolving, this series is particularly open to trans-disciplinary topics between fundamental science and engineering.

Indexed by EI-Compendex, SCOPUS and Springerlink.

More information about this series at https://link.springer.com/bookseries/8884

Eiji Tomita · Nobuyuki Kawahara ·
Ulugbek Azimov

Biogas Combustion Engines for Green Energy Generation

 Springer

Eiji Tomita (iD)
Okayama University
Okayama, Japan

Nobuyuki Kawahara (iD)
Department of Advanced Mechanics
Okayama University
Okayama, Japan

Ulugbek Azimov (iD)
Department of Mechanical
and Construction Engineering
Northumbria University
Newcastle upon Tyne, UK

ISSN 2191-530X   ISSN 2191-5318 (electronic)
SpringerBriefs in Applied Sciences and Technology
ISBN 978-3-030-94537-4   ISBN 978-3-030-94538-1 (eBook)
https://doi.org/10.1007/978-3-030-94538-1

This Springer imprint is published by the registered company Springer Nature Switzerland AG
The registered company address is: Gewerbestrasse 11, 6330 Cham, Switzerland

Preface

This book covers fundamentals on the combustion and exhaust gas emissions of biogas-fueled gas engines. Today, internal combustion engines for power generation are very common, using fossil fuels such as gasoline, diesel fuel, and natural gas. However, the carbon dioxide emitted from these engines will be limited in the future to prevent global warming. Therefore, we need to use green fuels in the future. Biogas is considered to be a renewable source of energy. Biogas can be produced by anaerobic digestion process from raw materials such as agricultural waste, manure, municipal waste, plant material, sewage, raw garbage, food waste, etc. It can also be used as a renewable energy source for the future, and will be very effective in helping to prevent global warming.

Biogas is a mixture of gases consisting mainly of methane and carbon dioxide, the ratio of which depends on the feedstock used for anaerobic digestion and the method of upgrading. The lower the percentage of carbon dioxide, the closer the fuel's properties are to natural gas. Internal combustion engines will never be gradually discontinued, on the contrary, they will be viable in the future as long as green fuels are used. In terms of energy conversion, engine systems have tremendous potential if they can be used for both the generation of electricity and production of heat from chemical energy sources.

The purpose of this book is to introduce to a reader the existing research on combustion and exhaust gas emissions for both biogas spark ignition engines and biogas dual-fuel engines. The book also describes several advanced combustion techniques to improve the thermal efficiency of engines, which could be directly applied to biogas engines. We hope that this book will be of interest to scientific researchers, professors, graduate students, and practicing engineers.

Okayama, Japan
Okayama, Japan
Newcastle upon Tyne, UK

Eiji Tomita
Nobuyuki Kawahara
Ulugbek Azimov

Contents

Chapter 1
Significance of Biogas, Its Production and Utilization in Gas Engines

Abstract Fighting global warming is urgent problem in the world now. This chapter describes the significance of preventing global warming. Effective utilization of biomass is one of the very important pathways for renewable energies. Among biofuels such as biochar, biodiesel, ethanol, producer gas, biogas, etc. this book focuses on biogas. Biogas is produced through anaerobic digestion (AD) process. Purified and upgraded biogas is used for producing high quality methane for injection in natural gas pipelines and engines. Biogas has been used in both, spark ignition engines and dual-fuel engines. It is important to utilize biogas engines for producing electricity and heat because biogas is suitable to local production for local consumption. Internal combustion engines are considered as important systems for conversion of chemical energy contained in biogas to power and electricity. The engine technologies have been developed and improved for a century. This book covers characteristics of combustion and exhaust emissions in spark ignition engines and dual-fuel engines fueled with biogas, as well as several new engine combustion technologies which can be applied to biogas engines.

Keywords Biogas · Anaerobic digestion · Global warming · Carbon dioxide · Methane · Gas engine · Purification of biogas · Upgrading of biogas · Greenhouse gas

Abbreviations

AD	Anaerobic digestion
CHP	Combined heat and power
CH_4	Methane
CO_2	Carbon dioxide
GHG	Greenhouse gas
GWP	Global warming potential
H_2S	Hydrogen sulfide
IPCC	Intergovernmental panel on climate change

© The Author(s), under exclusive license to Springer Nature Switzerland AG 2022   1
E. Tomita et al., *Biogas Combustion Engines for Green Energy Generation*,
SpringerBriefs in Applied Sciences and Technology,
https://doi.org/10.1007/978-3-030-94538-1_1

MCS   Microcrystalline silica
NH₃   Ammonia
NOₓ   Nitric oxides
PSA   Pressure swing adsorption
VFAs   Volatile fatty acids
VOCs   Volatile organic compounds
WBA   World Biogas Association

1.1   Significance of Preventing Global Warming

Scientists have been warning about climate change on Earth for decades. The Paris Agreement, a legally enforceable international climate change pact, was established in 2015. By the end of the twenty-first century, this accord suggests that when compared to pre-industrial levels, we must keep the change of world average temperatures below 2 °C, preferably 1.5 °C. According to the Intergovernmental Panel on Climate Change (IPCC) which stipulates that to achieve net zero global greenhouse gas (GHG) emissions by 2050 it is required to keep global temperatures below 1.5 °C [1]. It was recommended that by 2030, a sharp and persistent downward trajectory leading to a 45% reduction in carbon dioxide (CO_2) emissions from 2010 levels must be accomplished. Each country is now trying to reset a scenario to achieve the target for the year 2030. Increase in renewable energy is an urgent issue for every country. There are several renewable energies such as solar, wind, hydraulic, geothermal, tidal, etc., which are used for generating electricity. Biomass is also considered to be one of the important renewable energy sources.

The main gases which contribute to the greenhouse gas include about 76% CO_2 and about 16% methane (CH_4) (carbon dioxide equivalent) [2]. The ratio of CO_2 has increased gradually from 1970. In particular, CO_2 emissions due to combustion of fossil fuels and utilization of industrial processes were ~ 15 Gt/year (55%) in 1970 while they were ~ 32 GT/year (67%) in 2010. The reduction of carbon dioxide has received a great deal of attention because of the increase due to human activity of consumption of fossil fuels in the world. It is important to ensure a gradual decrease in fossil fuel combustion and consumption by power plants and internal combustion engines. However, the reduction of CO_2 emission becomes urgent now and it must be reduced quite significantly. Therefore, the measures to avoid burning fossil fuels such as coal, petroleum, and natural gas is discussed in industry section. In particular, a number of coal power plants will have to be decreased because of their very high CO_2 emissions per unit of power they produce.

According to the Fifth Assessment Report of the Intergovernmental Panel on Climate Change, the global warming potential (GWP) of CH_4, which is commonly expressed in terms of a 100-year timeframe, is roughly 28 times larger than that of CO_2 [3]. Over a 20-year period, however, the GWP of CH_4 becomes 85 times worse than that of CO_2 [3]. Therefore, lowering CH_4 emissions is also very critical.

1.2 Biomass and Biofuels

In IRENA's 1.5 °C scenario, bioenergy, including solid biomass, biogas, biomethane, and liquid biofuels, would account for 25% of total primary energy supply by 2050, according to Sect. 2.2.3 in IRENA 2021 [4], which equates to just over 150 EJ of biomass primary supply, a threefold increase over 2018 levels. The 1.5 °C climate target cannot be met without increased biomass production and usage. Biomass will be required in all aspects of the energy systems. It would play a major role in several industries, notably as a feedstock and fuel in the chemical industry, and as a fuel in the aviation industry.

Furthermore, negative emissions might be accomplished toward the net zero objectives if carbon capture and storage (CCS) technology is combined with biofuel production in the electricity sector and in some selected industrial sectors. There is a pressing need to produce more environmentally friendly fuels. As a renewable energy source, biomass is processed and categorized into solid, liquid, and gas. We can make solid (biochar), liquid (biodiesel, bio-alcohol), and gaseous fuels from plentiful biomass that is sustainable, CO_2-conserving, and commercially feasible (bio-hydrogen, syngas, biogas).

Biomass has features of (1) *renewable*: the balance of production and consumption is important, (2) *substitutive*: biomass can be used as alternative fuel of fossil fuel, (3) *storable*: biomass is stored as solid, liquid, and gaseous conditions unlike electricity, (4) *abundant*: there is a huge amount of biomass available on the Earth, and (5) *carbon neutral*: carbon in biomass is recycled during production and consumption.

The most essential approach to combat climate change is to convert energy and transportation systems as quickly as possible so that fossil carbon may be left in the ground. Sustainable bioenergy is now available and compatible with current energy infrastructure, allowing coal, natural gas, and petroleum fuels to be replaced immediately. Furthermore, when paired with carbon capture and storage associated with bioenergy usage, it can actually remove CO_2 from the environment. As a result, bioenergy can make a greater contribution in assisting energy vector transformation toward carbon neutrality.

There are three types of biomass power generation methods. (1) *Direct combustion method:* Water is boiled by the combusted biomass, and the steam is used to turn a turbine. The temperature that can be generated is low. Large equipment is required to generate electricity efficiently. (2) *Pyrolysis gasification method:* Biomass is steamed at high temperatures and the pyrolysis gas generated is used to generate electricity using a turbine or gas engine. Since the combustion temperature is high, power can be generated efficiently without the need for large equipment as in the direct combustion method. (3) *Biochemical gasification method:* Fermentation of biomass and use of the "biogas" generated in the process. Electricity can be generated efficiently. Even biomass that is difficult to burn can be used effectively. Various types of waste can be used effectively. In this textbook, biogas produced from anaerobic digestion is going to be utilized in gas engines.

As previously stated, methane in biogas has 28 times stronger negative impact as a greenhouse gas than carbon dioxide [3]. As a result, uncontained landfill gas that escapes into the atmosphere may play a substantial role in global warming. In addition, the production of photochemical smog is aided by the presence of volatile organic compounds (VOCs) which are emitted from landfill gas.

1.3 Sources of Biogas

Biogas is generated by microorganisms that undertake anaerobic respiration, such as methanogens and sulfate-reducing bacteria, in a natural or industrial settings. Biomethane is collected from industrial biogas and used as a fuel. The fermentation and anaerobic digestion of biological waste, organic fertilizers, biodegradable materials, sludge, sewage, trash, and energy crops, among other things, are used to produce biogas.

Large-scale, farm-based and commercial biogas facilities for power and heat are the emphasis of biogas development in developed nations. Furthermore, by upgrading biogas, high-quality biogas, mostly composed of methane, is utilized as a car fuel and for injection into the natural gas pipeline systems. Biogas was mostly produced in small, domestic-scale digesters in developing nations to offer a fuel for cooking or even illumination. Modern biogas facilities, on the other hand, are designed to generate energy while also providing heat [5].

Titled "Biogas: Pathways to 2030", a thorough study of how management of unavoidable organic waste through anaerobic digestion might achieve a 10% reduction in global GHG emissions—especially methane—by 2030, was published by the World Biogas Association (WBA) in 2021 [6].

There are various forms of biogas resources of substrates that are classified into three primary sources, namely, industrial, agricultural, and community wastes. The industrial wastes include municipal solid waste, industrial wastewater, etc. The agricultural wastes include animal manure, cereals, fruit waste, crop waste, harvest residues, etc. The community wastes include food wastes, wastewater sludge, etc. [7, 8].

Wet organic waste decomposes under anaerobic conditions to produce landfill gas, which is comparable to biogas. The weight of the material poured above covers the waste and compresses it mechanically. This substance protects anaerobic bacteria from being exposed to oxygen, allowing them to grow. If the facility is not designed to catch the gas, it builds up and is slowly discharged into the atmosphere. Dump gas discharged in unregulated manner can be dangerous because when it escapes from the landfill and combines with oxygen, it may have a risk of explosion. Food waste, in particular, accounts for not just resource loss but also land, energy, and water, with concurrent production of GHG emissions, which are a major contribution to global warming. Food that would not be properly used has negative health and environmental consequences. As a result, source separation and appropriate AD reduce the transmission of illnesses and smells, supporting healthy and hygienic sanitation.

Over 100 billion tons of organic wastes are produced each year [6]. Organic wastes are derived from a huge variety of sources: sewage sludge, breweries, slaughterhouses, bakeries, farms, dairy processors, restaurants, canteens, shops, bars, households, and so on. They are commonly classified into three categories, that is, *food waste*, *wastewater*, and *agricultural feedstocks*. Untreated organic waste is a problem. The AD system not only solves this problem, but also delivers the greatest value from these bioresources.

According to the report published by the World Biomass Association, the followings are estimated [6]: when good food is wasted, all of the land, energy, water and nutrients used to produce it are also wasted. Approximately, one third of all food produced is wasted, an estimated 1.3 billion tons worldwide. Halving this waste production could cut global emissions by 1650 tons of CO_{2e}, that is, 3% of total emissions. Of the wastes that cannot be avoided, from vegetable peels to sewage sludge, government must develop a comprehensive strategy to treat these wastes. By treating societies' unavoidable organic wastes, preventing these methane emissions, global GHG emissions would be cut by 5% each year. By placing AD at the heart of a circular economy of organic wastes, these organic "wastes" are converted into valuable "bioresources", namely biomethane, bio-CO_2, and bio-fertilizers. These products can help displace the need and use of traditional fossil-based alternatives, natural gas, fossil-derived CO_2 and energy-intensive artificial fertilizer. Using AD's sustainable products can cut global emissions by a further 5%. Therefore, total GHG savings from effective waste management is 13% while only 2% of these organics are currently being recycled. Therefore, it is expected to use the AD system widely in the world.

1.4 Anaerobic Digestion (AD)

Biogas technologies of production, yield, storage, upstream and downstream of anaerobic digestion (AD), management, utilization of biogas etc. are reviewed in some textbooks (e.g. [9–11]). In the absence of oxygen, anaerobic digestion has been proposed as a biodegradation technique for large-scale treatment of organic wastes. There have been published many review papers and book chapters about AD [8, 12–17], particularly for crop materials [18], sugarcane vinasse [19], municipal solid waste [20], biochar for enhancing biogas production of food waste and sludge [21], improving production using additives [22, 23], sludge pretreatment methods and co-digestion to boost biogas production and energy self-sufficiency in wastewater treatment plants [24], organic fraction of municipal solid waste [8], biowastes [25], upstream strategies of biological innovations to improve biogas production [7], new application processes for AD, such as phosphorus recovery, microbial fuel cells and seaweed digestion [26], the progress made related to feedstock, system configuration and operational conditions [27], etc.

The goal of AD is to turn waste into two types of useful and necessary materials: biogas and high-quality crop fertilizer [19]. Further enhancement of CH_4 production in AD is reviewed and discussed [28]. Biogas is produced through four-step processes which usually include hydrolysis, acidogenesis, acetogenesis, and methanogenesis.

Hydrolysis: The first stage is the hydrolysis process where high molecular weight complex insoluble organic matter is usually degraded into simple soluble molecules [8]. Insoluble organic compounds such as lipids, carbohydrates, and proteins are decomposed into soluble organic compounds such as glucoses (sugars), long-chain fatty acids and amino acids, respectively by hydrolytic bacteria that release their enzyme extracellularly [14].

Acidogenesis: The second stage is fermentation stage where degradation of the soluble compounds occurs from hydrolysis to the mixture of low molecular weight volatile fatty acids (VFAs), alcohols, CO_2 and H_2. The acidogenic bacteria are the most abundant bacteria and highly active fermenters in AD [8]. Acetic acid is formed in this reaction that is a major substrate for methanogens [29].

Acetogenesis: Acetogenesis is the third step, in which acetogenic bacteria transform acidogenesis products into acetate, H_2, and CO_2 [20]. VFAs, alcohols, certain amino acids, and aromatic compounds are metabolized by syntrophic acetogenic bacteria into methanogenesis substrates such as acetate, H_2, and CO_2 [8]. Dehydrogenation is another name for this step. Methanogens, on the other hand, absorb this H_2 and convert it to CH_4 [29].

Methanogenesis: The fourth stage called methanogenesis is the final metabolic stage of the anaerobic process [29]. Methanogenesis bacteria contained in rumen fluid in high amount converts biogas through mainly two reaction types, namely, hydrogenotrophic methanogenesis and acetoclastic methanogenesis. In hydrogenotrophic methanogenesis reaction, the bacteria change CO_2 and H_2 into CH_4 and H_2O. On the other hand, in acetoclastic methanogenesis reaction, the bacteria change acetic acid to CH_4 and CO_2. The latter process produces about 70% of CH_4 in AD compared to the former one of 30% production, because H_2 is limited in anaerobic process [18, 20]. They are anaerobes, which means they can exist in an oxygen-free environment and are vulnerable to a little amount of oxygen. They are highly essential microorganisms that develop slowly and are sensitive to substrate changes. Moreover, in methylotrophic methanogenesis, the bacteria convert methanol into CH_4, CO_2 and H_2O although the contribution of this reaction is very small [16].

Raw biogas is typically the mixture of CH_4 (50–75%) and CO_2 (25–50%) with traces of other gases that depend on substrate source and processing, those components include hydrogen sulfide (H_2S) (0–5000 ppm), ammonia (NH_3) (0–500 ppm), siloxanes (0–50 mg/m^3), N_2 (0–5%) and H_2O (1–5%) [20]. Biogas and nutritional substrate are the two major products of AD, as previously stated. Digestate, a combination of water and organic fertilizer for soil, is produced because of AD process. It's a type of biofertilizer that may be used to replace inorganic fertilizers.

1.5 Purification and Upgrading of Biogas

There have been published many review papers about upgrading of biogas [29–37]. Biogas consists of CH_4 and CO_2 as well as small amount of hydrogen sulfide, water droplets, nitrogen, ammonia, oxygen etc. The concentration of the impurities is different among the characteristics of the feedstock [38]. Biogas upgrading methods that have been proven are primarily taken from the natural gas purification sector.

Currently, physico-chemical methods for the primary separation of CH_4 and CO_2 are commercially ready, comprising absorption, adsorption, and membrane separation processes. In addition, new methods based on cryogenic processes or chemical hydrogenation are being developed.

The elimination of hazardous and/or poisonous components such as H_2S, NH_3, volatile organic compounds (VOCs), siloxanes, and moisture is the first stage in the treatment of biogas [36]. Due to their conversion to sulfur dioxide (SO_2) and sulfuric acid, many widely utilized gas applications, such as boilers, CHP (combined heat and power) engines, automobiles, or injection into the natural gas system, require the removal of water, H_2S, and other potential contaminants such as sulfuric acid (H_2SO_4). Ammonia (NH_3) is highly corrosive when combusted due to the conversion of nitric acid (HNO_3) and can be transformed into N_2O in nitric oxides (NO_x), which are part of GHG and polluting the environment. The removal of CO_2 is required for the fuel of vehicles and injection to grid because of the fuel density while the presence of CO_2 is allowed for CHP engines.

H_2S is corrosive and toxic that can cause significant harm to equipment, instruments, and piping. Several techniques are applied to remove H_2S during digestion and after digestion [30]. During digestion, there are two ways for removing H_2S. The sulfide is either oxidized to elemental sulfur or interacts with a metal ion to create an insoluble metal sulfide. (1) The biological aerobic oxidation of H_2S to elemental sulfur S by a group of specialized bacteria provides the basis for air or oxygen dosing; (2) Dosing iron chloride directly into the digester or into the influent mixing tank, resulting in FeS particle precipitation. After the digestion, H_2S is removed with some techniques. Scrubbing process is often applied with water or organic solvent. Adding chemicals in water can improve the absorption process. Diluted NaOH, $FeCl_2$ and $Fe(OH)_3$ are used as chemical absorption, forming Na_2S or NaHS, FeS and Fe_2S_3, respectively. Adsorption using iron oxide/hydroxides $Fe_2O_3/(Fe(OH)_3)$ forms FeS. Adsorption on activated carbon is also used to catalyze H_2S oxidization to elemental sulfur S. Membrane separation is also considered and reviewed [39–42].

A siloxane is a functional group in organosilicon chemistry with the $Si-O$ bonds. The biogas coming from landfill or waste composting causes this problem because siloxanes are widely used in industry as shampoos, detergents, cosmetics, pharmaceuticals, paper coatings, etc. When biogas is burned as an energy source, the siloxanes generate microcrystalline silica (MCS). Because of its glass characteristics, MCS causes abrasion, defective spark plugs, overheating of sensitive engine parts due to coating, and overall degradation of all mechanical engine elements [43].

To remove impurities, organic solvents, cryogenic separation, adsorption on silica gel, activated carbon, or other techniques are employed [30, 44].

Water vapor can be removed in several ways based on physical separation of condensed water and chemical drying, for example, glycerol, adsorption with silica gel, refrigeration, molecular sieves or activated carbon [30]. Refrigeration that makes the dewpoint lower to 0.5 °C was proven to be the simplest way for removing excess water vapor. Chemical drying methods of adsorption with silica, alumina or zeolites/molecular sieves are applied at elevated pressure. Absorption methods of water in tri-ethylene glycol or water with hygroscopic salt are also used.

Methods such as physical and chemical absorption, membrane separation, cryogenic separation, pressure swing adsorption (PSA), biological methane enrichment, and other techniques used to attain natural gas quality could be applied to biogas cleaning. The easiest, most environmentally friendly, cost-effective, and widely utilized approach for biogas cleaning and upgrading is physical absorption using water scrubbing [35]. The separation of H_2S and CO_2 from biogas is carried out because H_2S and CO_2 are significantly more soluble in water than CH_4. While water is flushed into the top of the tower, raw biogas is compressed to about 1 MPa and introduced to the bottom of the cleaning tower [45]. The upgraded biogas is taken from the top of the scrubbing column, and dissolved gases in the pressured water flowing from the bottom are desorbed in a flash tank or stripper at a pressure of 0.2–0.4 MPa.

Upgraded biogas is dried and compressed for storage and further transportation. The organic scrubbing uses organic solvents of polyethylene glycol, etc. to absorb CO_2 from biogas. Amines are used as chemical absorption method. PSA is carried out in vertical columns packed with absorbents, and the molecular sieves material is regenerated after adsorption, depressurization, desorption, and pressurization sequences [34]. CO_2 is absorbed while CH_4-rich gas passes via a pressured column. Adsorbents include silica gel, activated charcoal, activated carbon, synthetic resins, and zeolite. Following CO_2 saturation, biogas is sent to a fresh column, while the CO_2 saturated column is gradually depressurized to release a CO_2/CH_4 mixture with a high CH_4 concentration, which is vacuumed and returned to the PSA input. The membranes in biogas plants must withstand high pressure and harsh process conditions at the presence of H_2S and H_2O. Polymeric membrane is used because the material must be chemically resistant to those gases and withstand pressures of more than 25 bar and temperatures of more than 50 °C [41]. Cryogenic separation that uses the difference of boiling temperature of CO_2 (-78 °C) and CH_4 (-160 °C) is considered to separate CO_2. Liquefied biomethane (LBM) can be also produced and used instead of liquefied natural gas (LNG) [34]. Biological methane enrichment is the microbial conversion of H_2 and CO_2 into CH_4 driven by the capacity of hydrogenotrophic methanogens to utilize CO_2 as a carbon source and acceptor of electrons in the energy-producing process, and H_2 as an electron donor. Biological hydrogen methanation is reviewed [46], which would be a promising future utilization technique when solar energy increases significantly.

Processes, setups, bottlenecks, and efficiency of CO_2 bioconversion biotechnologies such as chemoautotrophic reactor based on H_2-assisted processes, gas fermentation, microbial electrochemical cells (MEC), and microalgae strain-based photosynthetic technique have lately been thoroughly investigated [47].

1.6 Biogas Utilization in Gas Engines

As described in Sect. 1.3, over 100 billion tons of organic wastes are produced each year. Without effective management, these rotting wastes can pollute the Earth. They contaminate drinking water; they reduce air quality; and they emit methane directly into the atmosphere. Methane is far more potent than carbon dioxide for GHG. The use of biogas in internal combustion engines integrated into the CHP systems can allow transforming the chemical energy of methane into both electrical power (approximately 30% ~ 40% and more in future) and heating and hot water (approximately 40%).

While electricity can be used advantageously for many uses, the degree of heat use in many cases is limited. Therefore, the total efficiency of CHPs often results in mediocre levels. Nonetheless, the use of biogas in CHPs is the most used, and a variety of commercial vendors offer standardized equipment. The increase in thermal efficiency of the engine is one of the issues for future CHPs.

Therefore, the effective usage of biogas for internal combustion engines as an alternative fuel is very important both for reducing the consumption of fossil fuel and for reducing the emission of methane itself. The amount of biomass material for biogas production is limited, however, the effective utilization processed through anaerobic digestion is one of the strongest candidates for reducing GHG in the world. By combining with other renewable energy technologies such as solar and wind power production, etc., net zero GHG emissions by 2050 is expected to be resolved. Internal combustion engines have been under development even now and considered to be superior devices of transforming energy from chemical material to power or electricity, even if fossil fuels are not used. This textbook describes the biogas combustion in internal combustion engines in Chaps. 2and3, and new combustion technologies in Chap. 4, which have been developed mainly in gasoline and diesel engines but can be easily applicable to biogas engines.

References

1. V. Masson-Delmotte, P. Zhai, H.O. Pörtner, D. Roberts, J. Skea, P.R. Shukla, A. Pirani, W. Moufouma-Okia, C. Péan, R. Pidcock, S. Connors, J.B.R. Matthews, Y. Chen, X. Zhou, M.I. Gomis, E. Lonnoy, T. Maycock, M. Tignor, T. Waterfield (eds.), Global Warming of 1.5°C. An IPCC Special Report on the impacts of global warming of 1.5°C above pre-industrial levels and related global greenhouse gas emission pathways, in the context of strengthening the global response to the threat of climate change, sustainable development, and efforts to eradicate poverty. Intergov. Panel Clim. Change (IPCC) (2018). https://www.ipcc.ch/sr15/. Accessed 20 Sept 2021

2. O. Edenhofer, R. Pichs-Madruga, Y. Sokona, E. Farahani, S. Kadner, K. Seyboth, A. Adler, I. Baum, S. Brunner, P. Eickemeier, B. Kriemann, J. Savolainen, S. Schlömer, C. von Stechow, T. Zwickel, J.C. Minx (eds.), IPCC: summary for policymakers, in *Climate Change 2014: Mitigation of Climate Change. Contribution of Working Group III to the Fifth Assessment Report of the Intergovernmental Panel on Climate Change* (Cambridge Univ Press, Cambridge, United Kingdom & New York, 2014). https://www.ipcc.ch/site/assets/uploads/2018/02/ipcc_wg3_ar5_summary-for-policymakers.pdf. Accessed 20 Sept 2021
3. G. Myhre, D. Shindell, F.M. Bréon, W. Collins, J. Fuglestvedt, J. Huang, D. Koch, J.F. Lamarque, D. Lee, B. Mendoza, T. Nakajima, A. Robock, G. Stephens, T. Takemura, H. Zhang, Anthropogenic and natural radiative forcing, in *Climate Change 2013: The Physical Science Basis. Contribution of Working Group I to the Fifth Assessment Report of the Intergovernmental Panel on Climate Change*, ed. by T.F. Stocker, D Qin, G.K. Plattner, M. Tignor, S.K. Allen, J. Boschung, A. Nauels, Y. Xia, V. Bex, P.M. Midgley (Cambridge University Press, Cambridge, United Kingdom & New York, 2013). https://www.ipcc.ch/site/assets/uploads/2018/02/WG1 AR5_Chapter08_FINAL.pdf. Accessed 20 Sept 2021
4. IRENA, *World Energy Transitions Outlook: 1.5°C Pathway* (International Renewable Energy Agency, Abu Dhabi, 2021). https://www.irena.org/publications. Accessed 20 Sept 2021
5. N. Scarlat, J.F. Dallemand, F. Fahl, Biogas: developments and perspectives in Europe. Renew. Energy **129A**, 457–472 (2018). https://doi.org/10.1016/j.renene.2018.03.006
6. WBA, *Biogas: Pathways to 2030* (World Biogas Association (WBA), 2021). https://www.worldbiogasassociation.org/biogas-pathways-to-2030-report/. Accessed 20 Sept 2021
7. M. Tabatabaei, M. Aghbashlo, E. Valijanian, H. Kazemi Shariat Panahi, A.S. Nizami, H. Ghanavati, A. Sulaiman, S. Mirmohamadsadeghi, K. Karimi, A comprehensive review on recent biological innovations to improve biogas production, Part 1: Upstream strategies. Renew. Energy 146, 1204–1220 (2020). https://doi.org/10.1016/j.renene.2019.07.037
8. M.F.M.A. Zamri, S. Hasmady, A. Akhiar, F. Ideris, A.H. Shamsuddin, M. Mofijur, I.M. Rizwanul Fattah, T.M.I. Mahlia, A comprehensive review on anaerobic digestion of organic fraction of municipal solid waste. Renew. Sustain. Energy Rev. **137**, 110637 (2021). https://doi.org/10.1016/j.rser.2020.110637
9. M. Tabatabaei, H. Ghanavati (eds.), *Biogas—Fundamentals, Process, and Operation* (Springer International Publishing, 2018)
10. H. Treichel, G. Fongaro (eds.), *Improving Biogas Production—Technological Challenges, Alternative Sources, Future Developments* (Springer Int. Publishing, 2019)
11. L. Deng, Y. Liu, W. Wang (eds.), *Biogas Technology* (Springer, Singapore, 2020)
12. P. Weiland, Biogas production: current state and perspectives. Appl. Microbiol. Biotechnol. **85**, 849–860 (2010). https://doi.org/10.1007/s00253-009-2246-7
13. K. Ziemiński, M. Frąc, Methane fermentation process as anaerobic digestion of biomass: Transformations, stages and microorganisms. Afr. J. Biotechnol. **11**, 4127–4139 (2012). https://www.ajol.info/index.php/ajb/article/view/101067
14. B. Bharathiraja, T. Sudharsanaa, A. Bharghavi, J. Jayamuthunagai, R. Praveenkumar, Biohydrogen and biogas—an overview on feedstocks and enhancement process. Fuel **185**, 810–828 (2016). https://doi.org/10.1016/j.fuel.2016.08.030
15. M.A. Mir, A. Hussain, C. Verma, Design considerations and operational performance of anaerobic digester: a review. Cogent. Eng. **3**, 1181696 (2016). https://doi.org/10.1080/23311916.2016.1181696
16. I. Syaichurrozi, R. Rusdi, T. Hidayat, A. Bustomi, Kinetics Studies Impact of Initial pH and Addition of Yeast Saccharomyces cerevisiae on Biogas Production from Tofu Wastewater in Indonesia. Int. J. Eng. Trans. B: Appl. **29**, 1037–1046 (2016). https://www.ije.ir/article_72765.html
17. K. Obileke, N. Nwokolo, G. Makaka, P. Mukumba, H. Onyeaka, Anaerobic digestion: technology for biogas production as a source of renewable energy—a review. Energy Environ. **32**, 191–225 (2020). https://doi.org/10.1177/0958305X20923117
18. R. Braun, Anaerobic digestion: a multi-faceted process for energy, environmental management and rural development, in *Improvement of Crop Plants for Industrial End Use*, ed. by P. Ranalli (Springer, Netherlands, 2007). https://doi.org/10.1007/978-1-4020-5486-0_13

19. M. Parsaee, M. Kiani Deh Kiania, K. Karimi, A review of biogas production from sugarcane vinasse. Biomass Bioenergy **122**, 117–125 (2019). https://doi.org/10.1016/j.biombioe.2019.01.034

20. E.N. Richard, A. Hilonga, R.L. Machunda, K.N. Njau, A review on strategies to optimize metabolic stages of anaerobic digestion of municipal solid wastes towards enhanced resources recovery. Sustain. Environ. Res. **29**, 36 (2019). https://doi.org/10.1186/s42834-019-0037-0

21. M. Kumar, S. Dutta, S. You, G. Luo, S. Zhang, P.L. Show, A.D. Sawarkar, L. Singh, D.C.W. Tsang, A critical review on biochar for enhancing biogas production from anaerobic digestion of food waste and sludge. J. Clean. Prod. **305**, 127143 (2021). https://doi.org/10.1016/j.jclepro.2021.127143

22. M.S. Romero-, J. Vila, J. Mata-Alvarez, J.M. Chimenos, S. Astals, The role of addi tives on anaerobic digestion: a review. Renew. Sustain. Energy Rev. **58**, 1486–1499 (2016). https://doi.org/10.1016/j.rser.2015.12.094

23. M. Liu, Y. Wei, X. Leng, Improving biogas production using additives in anaerobic digestion: a review. J. Clean. Prod. **297**, 126666 (2021). https://doi.org/10.1016/j.jclepro.2021.126666

24. I. Volschan Junior, R. de Almeida, M.C. Cammarota, A review of sludge pretreatment methods and co-digestion to boost biogas production and energy self-sufficiency in wastewater treatment plants. J. Water Process. Eng. **40**, 101857 (2021). https://doi.org/10.1016/j.jwpe.2020.101857

25. S. Achinas, V. Achinas, G.J.W. Euverink, A technological overview of biogas production from biowaste. Engineering **3**, 299–307 (2017). https://doi.org/10.1016/J.ENG.2017.03.002

26. N. Horan, A.Z. Yaser, N. Wid (eds.), *Anaerobic Digestion Processes—Applications and Effluent Treatment* (Springer, Singapore, 2018). https://doi.org/10.1007/978-981-10-8129-3

27. N. Balagurusamy, A.K. Chandel (eds.) *Biogas Production—From Anaerobic Digestion to a Sustainable Bioenergy Industry* (Springer Int. Publishing, 2020). https://doi.org/10.1007/978-3-030-58827-4

28. Y. Li, Y. Chen, J. Wu, Enhancement of methane production in anaerobic digestion process: A review. Appl. Energy **240**, 120–137 (2019). https://doi.org/10.1016/j.apenergy.2019.01.243

29. D. Thiruselvi, P.S. Kumar, M.A. Kumar, C.H. Lay, S. Aathika, Y. Mani, D. Jagadiswary, A. Dhanasekaran, P. Shanmugam, S. Sivanesan, P.L. Show, A critical review on global trends in biogas scenario with its up-gradation techniques for fuel cell and future perspectives. Int. J. Hydrogen Energy **46**, 16734–16750 (2021). https://doi.org/10.1016/j.ijhydene.2020.10.023

30. E. Ryckebosch, M. Drouillon, H. Vervaeren, Techniques for transformation of biogas to biomethane. Biomass Bioenergy **35**, 1633–1645 (2011). https://doi.org/10.1016/j.biombioe.2011.02.033

31. F. Bauer, T. Persson, C. Hulteberg, D. Tamm, Biogas upgrading—technology overview, comparison and perspectives for the future. Biofuels Bioproducts Biorefining **7**, 499–511 (2013). https://doi.org/10.1002/bbb.1423

32. R. Muñoz, L. Meier, I. Diaz, D. Jeison, A review on the state-of-the art of physical/chemical and biological technologies for biogas upgrading. Rev. Environ. Sci. Biotechnol. **14**, 727–759 (2015). https://doi.org/10.1007/s11157-015-9379-1

33. Q. Sun, H. Li, J. Yan, L. Liu, Z. Yu, X. Yu, Selection of appropriate biogas upgrading technology-a review of biogas cleaning, upgrading and utilization. Renew. Sustain. Energy Rev. **51**, 521–532 (2015). https://doi.org/10.1016/j.rser.2015.06.029

34. O.W. Awe, Y. Zhao, A. Nzihou, D.P. Minh, N. Lyczko, A review of biogas utilisation, purification and upgrading technologies. Waste Biomass Valor. **8**, 267–283 (2017). https://doi.org/10.1007/s12649-016-9826-4

35. I. Angelidaki, L. Treu, P. Tsapekos, G. Luo, S. Campanaro, H. Wenzel, P.G. Kougiasa, Biogas upgrading and utilization: current status and perspectives. Biotechnol. Adv. **36**, 452–466 (2018). https://doi.org/10.1016/j.biotechadv.2018.01.011

36. R. Kapoor, P. Ghosh, M. Kumar, V.K. Vijay, Evaluation of biogas upgrading technologies and future perspectives: a review. Environ. Sci. Pollut. Res. **26**, 11631–11661 (2019). https://doi.org/10.1007/s11356-019-04767-1

37. E. Mulu, M.M. M'Arimi, R.C. Ramkat, A review of recent development in application of low cost natural materials in purification and upgrade of biogas. Renew. Sustain. Energy Rev. **145**, 111081 (2021). https://doi.org/10.1016/j.rser.2021.111081

38. N.I.H.A. Aziz, M.M. Hanafiaha, S.H. Gheewala, A review on life cycle assessment of biogas production: challenges and future perspectives in Malaysia. Biomass Bioenergy **122**, 361–374 (2019). https://doi.org/10.1016/j.biombioe.2019.01.047

39. T.E. Rufford, S. Smart, G.C.Y. Watson, B.F. Graham, J. Boxall, J.C. Diniz, E.F. May, The removal of CO_2 and N_2 from natural gas: a review of conventional and emerging process technologies. J. Pet. Sci. Eng. **94–95**, 23–154 (2012). https://doi.org/10.1016/j.petrol.2012.06.016

40. J.K. Adewole, A.L. Ahmad, S. Ismail, C.P. Leo, Current challenges in membrane separation of CO2 from natural gas: a review. Int. J. Greenh. Gas Con. **17**, 46–65 (2013). https://doi.org/10.1016/j.ijggc.2013.04.012

41. M. Scholz, T. Melin, M. Wessling, Transforming biogas into biomethane using membrane technology. Renew. Sustain. Energy Rev. **17**, 199–212 (2013). https://doi.org/10.1016/j.rser.2012.08.009

42. I.U. Khan, M.H. Othman, H. Hashim, T. Matsuura, A.F. Ismail, M. Rezaei-Dasht Arzhandi, I.W. Azelee, Biogas as a renewable energy fuel—a review of biogas upgrading, utilisation and storage. Energy Convers. Manage. **150**, 277–294 (2017). https://doi.org/10.1016/j.enconman.2017.08.035

43. N. Abatzoglou, S. Boivin, A review of biogas purification processes. Biofuels Bioprod. Biorefin. **3**, 42–71 (2009). https://doi.org/10.1002/bbb.117

44. M. Ajhar, M. Travesset, S. Yüce, T. Melin, Siloxane removal from landfill and digester gas—a technology overview. Bioresour. Technol. **101**, 2913–2923 (2010). https://doi.org/10.1016/j.biortech.2009.12.018

45. S. Sahota, G. Shah, P. Ghosh, R. Kapoor, S. Sengupta, P. Singh, V. Vijay, A. Sahay, V.K. Vijay, I.S. Thakur, Review of trends in biogas upgradation technologies and future perspectives. Bioresour. Technol. Reports **1**, 79–88 (2018). https://doi.org/10.1016/j.biteb.2018.01.002

46. B. Lecker, L. Illi, A. Lemmer, H. Oechsner, Biological hydrogen methanation—a review. Bioresour. Technol. **245**, 1220–1228 (2017). https://doi.org/10.1016/j.biortech.2017.08.176

47. L. Wu, W. Wei, L. Song, M. Wózniak-Karczewska, Ł. Chrzanowski, B.J. Ni, Upgrading biogas produced in anaerobic digestion: biological removal and bioconversion of CO_2 in biogas. Renew. Sustain. Energy Rev. **150**, 111448 (2021). https://doi.org/10.1016/j.rser.2021.111448

Chapter 2
Combustion and Exhaust Emissions of Biogas Spark Ignition Engines

Abstract Biogas is mainly composed of methane (CH_4) and carbon dioxide (CO_2) as presented in Chap. 1. Spark ignition reciprocating engines are often used for biogas engine because of only small modifications of fuel supply system from gasoline engine especially at small output powers. At first, spark discharge, burning velocity and flame structure are explained. Next, the effects of CO_2 ratio, equivalence ratio, compression ratio, hydrogen addition, exhaust gas recirculation (EGR), fuel property and other physical parameters are shown based on the literature sources published. An example of analyzing method of combustion by using pressure history data is introduced combined with NO_x (oxides of nitrogen) emission data. When CO_2 ratio in the fuel increases, the burning duration becomes longer, and the burning rate becomes smaller due to smaller laminar burning velocity. Abnormal combustion of knock must be avoided in spark ignition engines. Autoignition and knock behavior in the end-gas region is visualized with a high-speed camera. Methane number as an indicator of resistance to knock is discussed. Pre-chamber with small holes, from where burned gas jets come out and make several ignition locations in a main chamber, is used in lean burn, large-sized bore engines because of shorter flame development time.

Keywords Spark ignition engine · Combustion · Exhaust emissions · Biogas · Burning velocity · Spark discharge · Autoignition · Knocking · Methane number · Exhaust gas recirculation

Abbreviations

AD	Anaerobic digestion
BSFC	Brake specific fuel consumption
CAD	Crank angle degree
CCR	Critical compression ratio
CFD	Computational fluid dynamics
CNG	Compressed natural gas

CO   Carbon monoxide
EGR   Exhaust gas recirculation
HC   Hydrocarbons
IMEP   Indicated mean effective pressure
LPG   Liquefied petroleum gas
MBT   Minimum advance for best torque (or maximum brake torque)
MFB   Mass fraction burned
MN   Methane number
NO_x Oxides of nitrogen
ROHR   Rate of heat release
TDC   Top dead center
T-PAD   Two-phase anaerobic digestion
WOT   Wide open throttle

2.1 Introduction

Biogas is one of the typical biofuels generated from biomass wastes through anaerobic digestion (AD). The biogas from conventional AD usually contains methane, CH_4 (50–75 vol%) and carbon dioxide, CO_2 (25–50 vol%), as shown in Chap. 1. Before using in internal combustion engines, impurities in biogas such as H_2S, should be removed to avoid corrosion problem [1, 2].

This chapter describes the combustion, engine performance and exhaust emissions in spark ignition biogas engines. At first, spark discharge, burning velocity and flame structure are described. Next, an example of combustion analysis is described in varying CO_2 to CH_4 ratio in the imitated biogas and equivalence ratio. Thereafter, the effects of compression ratio, addition of hydrogen, EGR, fuel properties, etc. on engine performance and exhaust emissions are described.

The effect of concentration of CO_2 on engine performance is investigated in many works because the ratio varies due to the production method and the feedstock material of raw biogas. The presence of CO_2 has a diluent effect that leads to the suppression of the combustion rate. Although the expense of removing CO_2 from biogas should be addressed, it is commonly done to enhance its heating value for the use in spark ignition engines. Hydrogen addition and increase in compression ratio etc. are the methods to recover the combustion intensity. However, too much hydrogen addition may increase NO_x (Oxides of nitrogen) and finally induce knocking. To reduce the combustion speed, EGR is often used in internal combustion engines.

Although raw biogas such as landfill gas etc. is used in real engines, simulated biogas composed of 60% CH_4 and 40% CO_2 is also used very often for fundamental studies. Fuel supply systems are mainly carburetors or fuel injection to intake port. Due to the vastly differing stoichiometric air-to-fuel requirements, carburetors designed for other gaseous fuels such as natural gas or liquefied petroleum gas

(LPG) are inappropriate. Based on the amount of methane in the gas, the stoichiometric air-to-fuel ratio varies between 10 and 6 (on a volume basis) for biogas/landfill gas fuels.

2.2 Fundamentals of Combustion in Spark Ignition Engines

2.2.1 Spark Discharge

In spark ignition engines, the initiation of combustion is due to electrical discharge. High voltage is supplied to the gap between spark electrode and ground of the spark plug near the end of the compression stroke. The transistorized coil ignition (TCI) system is often used because of unsurpassed benefit and cost. The electrical energy is stored in the inductance of a coil and released slowly over about 2 ms.

Figure 2.1 shows typical examples of voltage and current (upper), and energy and power (lower) of the ignition system as a function of time [3, 4]. The actual values depend on the electrical components of the discharge circuits. The ignition process is divided into three phases, such as breakdown, arc, and glow discharge. The breakdown phase is characterized by high voltage (~ 10 kV), high peak current (~ 200 A), and an extremely short duration (~ 10 ns). A cylindrical plasma channel (~ 40 μm in diameter) is formed, where the gas molecules are fully dissociated and ionized. The energy supplied is transferred almost without loss to the plasma column. The pressure and temperature rise very rapidly to the values up to 20 MPa and 60,000 K, respectively. This causes an intense shock wave with expanding the channel. Thus, the pressure and temperature decrease with time and the energy is reconverted to thermal energy. The energy loss about 30% occurs due to the shock wave. However, most of the energy regained by the plasma because spherical blast waves transfer most of their energy to the gas within a small spherical region.

The arc and the glow discharge follows after the breakdown phase. The conductive path is created between the electrodes. The arc voltage is very low (< 100 V), although the current is as high as external circuit permits. Although the degree of dissociation is quite high in the central region of the discharge, the degree of ionization is only one percent. The arc voltage drops at cathode and anode of metal electrodes are significant. As the arc requires a hot cathode spot, there is also severe erosion of the cathode material due to evaporation. The arc expands mainly due to heat conduction and mass diffusion. Due to continuous energy loss, the equilibrium kernel gas temperature is limited to about 6000 K. In glow discharge, the typical values drop to the values; current less than 200 mA, voltage at cathode about 300–500 V, and ionization less than 0.01%. Overall losses are higher than in the arc, so that the equilibrium gas temperature is about 3000 K.

The minimum energy for ignition of quiescent stoichiometric methane-air mixture at normal engine conditions due to a spark is about 0.3 mJ [5]. However, conventional ignition systems deliver 30–50 mJ of electrical energy to the spark for secure ignition because the minimum value increases in lean and/or strong gas flow conditions.

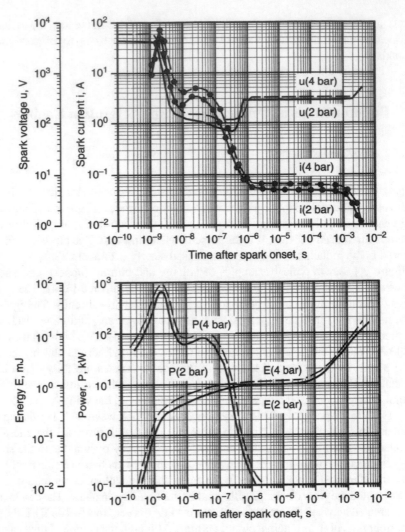

Fig. 2.1 Typical examples of voltage and current (upper), and its energy and power (lower) of the ignition system as a function of time [4]

2.2.2 Burning Velocity and Flame Structure

Laminar burning velocity is considered as one of the fundamental characteristics for determining combustion behavior of premixed mixtures. Review papers on laminar burning velocity for various fuel–air mixtures [6] and for flammability limit [7] were published. The definition of laminar burning velocity is the speed that unburned gas enters into the flame front perpendicularly. It is easy to understand for stabilized flame such as burner. For propagating flame in a combustion chamber, we must notice that the burning velocity is the apparent flame speed minus expansion velocity due to burned gas. The laminar burning velocities have been measured and simulated for biogas (methane-carbon dioxide) and air mixture for wide range of pressures and temperatures [8]. The laminar burning velocity for CH_4/CO_2 and air mixture for both experiment and simulation was investigated and presented that the hydrogen addition promotes the laminar burning velocity [9, 10]. The laminar burning velocity depends on ambient pressure, unburned gas temperature, gaseous fuel components, its equivalence ratio of fuel–air mixture and dilution ratio of EGR.

Figure 2.2 shows adiabatic laminar burning velocities of $CH_4/CO_2/air$ mixtures at 298 K and 1 atm for equivalence ratio ϕ between 0.7 and 1.4 [11]. Numerical results simulated by four chemical kinetics mechanisms such as San Diego [12], GRI-Mech 3.0 [13], USC Mech II [14], Konnov [15], and experimental data from two literature sources [16, 17] are also presented. All the data shows the same trend and almost the same values. The laminar burning velocity presents its maximum value near stoichiometric condition ($\phi = 1$). When the dilution ratio of carbon dioxide X_{CO2} increases from 0.1 to 0.5, the laminar burning velocity decreases. This is because the high heat capacity of CO_2 reduces the adiabatic flame temperature.

When the flame surface has curvature or the flame is distorted by the flow velocity gradient, the flame undergoes elongation and the burning velocity generally changes. The Markstein number characterizes the variation in the local burning velocity due to the effect of stretching. Using stretch rate α and Markstein length L, the relationship between the laminar burning velocity S_L when unstretched and the laminar burning velocity S_{L0} when stretched is derived [18].

$$S_L - S_{L0} = \alpha L \tag{2.1}$$

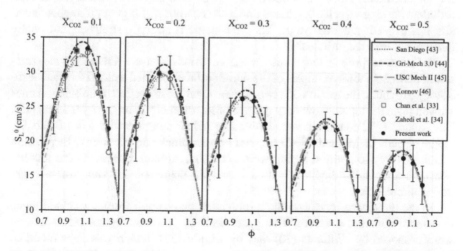

Fig. 2.2 Laminar burning velocity of CH_4/CO_2 and air mixture at 298 K and 1 atm (CO_2 ratio $X_{CO2} = 0.1$–0.5) compared to the results by Chan et al. [16], Zahedi et al. [17] as well as simulation results of San Diego [12], GRI-Mech [13], USC Mech II [14] and Konnov [15]. (Nonaka and Pereira [11])

The stretch rate α is expressed by

$$\alpha = (1/A)\mathrm{d}A/\mathrm{d}t \tag{2.2}$$

where A and t are an infinitesimal area of the flame surface and time, respectively. The Markstein number Ma is defined as

$$Ma = L/\delta_\mathrm{L} \tag{2.3}$$

where characteristic thickness of preheat zone of the flame δ_L is estimated by using the classical definition of $\delta_\mathrm{L} = \lambda_0 /(c_\mathrm{p}\rho_\mathrm{u}S_\mathrm{L})$ with λ_0 the thermal conductivity, c_p the specific heat capacity of unburned gas and ρ_u the unburned density.

The flame propagation speed depends on turbulent burning velocity S_T, which is a function of laminar burning velocity S_L and turbulence intensity u'. As the turbulence intensity increases, for example, the turbulent burning velocity increases as follows [4], although there were many equations proposed.

$$S_\mathrm{T}/S_\mathrm{L} = 1 + \left(u'/S_\mathrm{L}\right)^n \tag{2.4}$$

for unstretched case, where $n \approx 5/6 \dots 1$ and

$$S_\mathrm{T}/S_\mathrm{L} = I_0\left[1 + C_0\left(u'/S_\mathrm{L}\right)^n\right] \tag{2.5}$$

for stretched case, where $I_0 =$ strain rate, $C_0 \approx 1 \dots 2.5$ with straining. However, turbulent burning velocity decreased under the condition of higher turbulence intensity that exceeds a certain value. Then the flame is strongly stretched and finally quenched due to the turbulence.

The turbulent propagating flames in engine cylinders have wrinkled surface structure. Figure 2.3 shows examples of the flame surface obtained using laser sheet technology to capture the motion of a two-dimensional cross-section of turbulent flames viewed from the top window for a quarter of its travel direction (every 0.3 ms) [19]. The experimental conditions were iso-octane and air mixtures with $\phi = 1.0$ and 0.8, engine speeds of 750 and 1500 rpm, and boosted intake air pressure. The pressure conditions for the ignition timing were very high, around 3 MPa. As the mixture became leaner and/or engine speed increased, the degree of wrinkling was stronger due to turbulence.

As the turbulence increases, the turbulent flame changes from a wrinkled flame to a more complex structure. There are two most popular regime diagrams that were proposed by Williams [20] and by Borghi [21], which was later modified by Peters [22]. Figure 2.4 shows the latter one called as Peters-Borghi diagram, which is presented by parameters such as turbulence intensity q, integral scale L_T, laminar burning velocity S_L, and laminar flame thickness δ_L [22]. The structure of

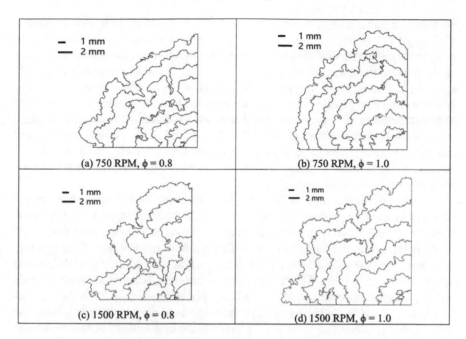

Fig. 2.3 Top view of turbulent flame development in 1/4 area at two different engine speeds (750 and 1500 rpm) and two different equivalence ratios (0.8 and 1.0) [19]

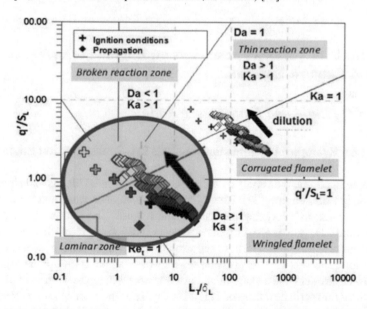

Fig. 2.4 Effect of EGR on flame structure transition in Peters-Borghi diagram (Red: 10%EGR, Green: 15%EGR, Purple: 20%EGR, Yellow: 25%EGR, White: 30%EGR) [23]

the turbulent flame is classified using Damköhler number Da and Karlovitz number K. Damköhler number is expressed as the ratio of characteristic eddy turnover time scale to chemical reaction time scale, $(L_T/q')/(\delta_L/S_L)$. Karlovitz number is defined as the ratio of the chemical time scale t_F ($=\delta_L/S_L$) to the Kolmogorov time scale t_η ($=\nu/\varepsilon)^{1/2}$, that is, (δ_L/S_L) $(\varepsilon/\nu)^{1/2}$, where ε and ν are the dissipation of turbulent kinetic energy and kinetic viscosity of the unburned gas, respectively. By using the Kolmogorov scale η, the parameter K is also written as δ_L^2/η^2. The laminar flame is located in the left and bottom side in Fig. 2.4. As the ratio of turbulence intensity to laminar burning velocity increases, the flame structure changes from wringled flamelet to broken reaction zone through corrugated flamelet and thin reaction zone according to the value of L_T/δ_L [22].

The effect of dilution rate on the flame structure transition is displayed in Fig. 2.4 [23]. The atmosphere conditions were pressure 1.5–4 MPa, temperature 600–800 K at engine speed of 1300 rpm. The crosses and diamonds indicate the conditions at the spark timing and at the flame conditions after ignition, respectively. After ignition, the flame structure shifted from right-down direction linearly for each EGR condition less than 25% EGR. This is because the turbulent burning velocity depends on the turbulence intensity as shown in (2.4). For the case with EGR 30% condition, however, a turn could be seen after the progress of the flame propagation. The researchers explained that there was no dependence between the turbulent burning velocity and the turbulence intensity because the smallest eddies had possibilities to penetrate into the preheated zone in thin reaction zone [23].

2.3 Effects of Various Parameters on Engine Performance and Exhaust Emissions

2.3.1 CO₂ Ratio and Equivalence Ratio

2.3.1.1 An Example of Combustion Analysis in a Spark-Ignition Engine

An example of combustion analysis is presented for simulated biogas and air mixture in a spark-ignition engine [24]. The effects of equivalence ratio and ratio of CO_2 were investigated by analyzing the rate of heat release. The relation between NO_x emissions and CH_4 ratio is also presented. A spark ignition engine with three cylinders for co-generation was used. The bore and stroke were 88 and 90 mm, respectively. The compression ratio was 12:1. A pressure sensor was adapted to the spark plug. Using mass-flow controllers and a static mixer, researchers studied the simulated biogas by altering the volumetric flow rates of CH_4 and CO_2. The equivalence ratio was adjusted with the use of a throttle valve. The system's highest electrical output power was 6 kW. The load generated by the engine was absorbed by ceramic heaters, which were controlled by changing the internal and external electric power outputs. The engine was operated at 1700 rpm, that allowed producing electricity continuously.

The spark was initiated at 20° before TDC (top dead center). The fuel mix was adjusted from 100% CH_4 to 60% CH_4 and 40% CO_2 by regulating the gas flow rate. The operation was examined at stoichiometric and lean burn conditions. The equivalence ratio was calculated without taking CO_2 flow into account, only using the intake air and methane flow rates alone.

Under the lean operating limits, the possible engine operating conditions were studied, as illustrated in Fig. 2.5. The engine operating regimes were studied for the range from stoichiometric to the lean conditions. The study also included the engine operating regime to maintain engine at 1700 rpm at lean conditions. Controlling the electrical output power instigated the change of the engine load throughout a range of 1.85–5.90 kW. The engine could be run at a lower equivalence ratio when the CH_4 ratio and engine load increased. This is due to the increase of mixture density and the decrease of the pumping losses when the throttle is opened wider at higher engine loads.

Four distinct engine load regimes were studied to see how the CO_2 ratio in the fuel affected combustion and NO_x formation. For the range of conditions at electrical power outputs at 1.85, 3.20, 4.55, and 5.90 kW, the equivalence ratio ϕ was varied from the stoichiometric and then adjusted as lean between 0.9, 0.8, 0.775, and 0.75,

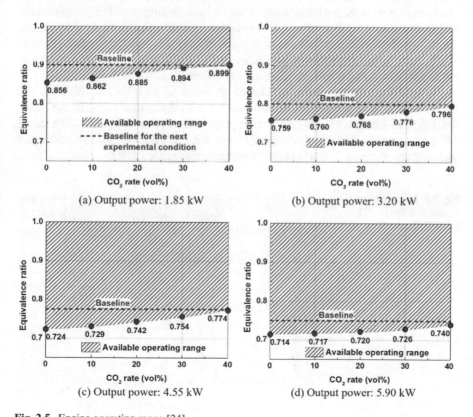

Fig. 2.5 Engine operating range [24]

Fig. 2.6 In-cylinder pressure and rate of heat release (ROHR) for output power of 5.9 kW ($\phi = 0.75$) [24]

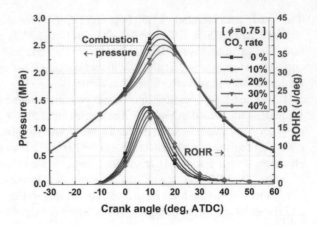

as shown in Fig. 2.5. This range was determined based on the results of each load approaching the lean limit. Figure 2.6 illustrates an example of in-cylinder pressure and the rate of heat release (ROHR) for various CO_2 ratios for a 5.9 kW engine output and a 0.75 equivalence ratio. When compared to stoichiometric fuel–air ratios, the combustion phases were extended and delayed for lean mixtures. The peak combustion pressure and ROHR fell as the CO_2 ratio rose, and the combustion slowed down. When considering the inert gas characteristics of CO_2, this can make sense. Because of the lower levels of maximum pressures and temperatures, combustion of biogas can result in lower combustion noise and NO_x formation.

Figure 2.7 shows the characteristics of the mass fraction burned (MFB) under lean condition ($\phi = 0.75$) with an output of 5.9 kW. By analyzing MFB, the initial (0–10%) and main (10–80%) combustion periods are shown in Fig. 2.8. The initial combustion time is linked to the longer ignition delay as the CO_2 ratio rose, as did the main combustion length. Both combustion durations reduced as the engine load rose;

Fig. 2.7 Mass fraction burned at different CO_2 ratios [24]

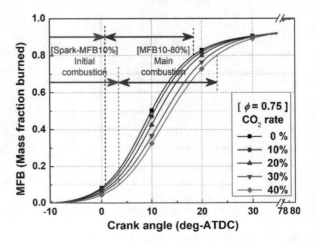

Fig. 2.8 Changes in combustion periods at different CO_2 and equivalence ratio conditions [24]

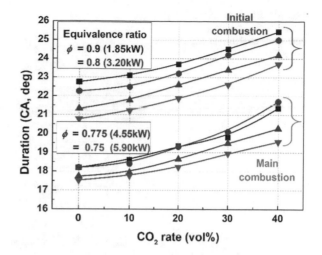

however, as the engine load further increased, the disparities in the primary combustion duration between 100% CH_4 and the fuel containing CO_2 gradually diminished and became identical at the greatest engine load. This is because, as the fuel–air ratio increases under greater engine loads, combustion becomes more vigorous, and the influence of the inert gas diminishes. Comparison of lean burn conditions to stoichiometric fuel–air mixtures is difficult because of the varied equivalence ratios for each engine load; nevertheless, under lean burn conditions, the initial and main combustion durations were often greater, and the combustion durations became longer with increasing CO_2 fraction.

Figure 2.9 depicts NO_x emissions for all engine loads as a function of CO_2 ratio. NO_x emissions fell considerably as the equivalence ratio decreased for all engine loads. The NO_x emissions surpassed the maximum limit of the measuring equipment at low CO_2 ratios when operating at stoichiometric fuel–air ratios and at higher engine loads. However, as the CO_2 ratio increased, NO_x emissions fell gradually. In general, a lean-burn strategy can minimize NO_x emissions for a given engine load, and a fuel containing CO_2 can further instigate the reduction of NO_x emissions. NO_x is produced primarily during combustion when the combination of nitrogen and oxygen is heated at high gas temperatures. Because of CO_2's high specific heat and inert nature, the maximum combustion temperature is reduced. The impact is comparable to the conditions of EGR that lowers combustion temperature.

To summarize, since the cost of biogas production with a higher methane content has been substantially reduced, as a result of recent technological advancements, the range of options has widened. The approaches which take into account both environmental issues and production costs are required to maximize benefits of using biogas in internal combustion engines.

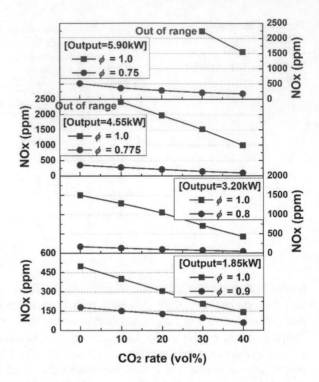

Fig. 2.9 Effect of CO_2 concentration on NO_x emissions at different engine loads [24]

2.3.1.2 *Other Research for Varying CO_2 Ratio*

The effect of ratio of CO_2 in the biogas has been investigated by many researchers. Simulated biogas of CH_4 and CO_2 is often used to operate an engine with the ratio of CO_2 0–50% in the fuel. When biogas composed of 65% CH_4, 35% CO_2 is used into a gasoline engine, biogas replaces some of the intake air [25]. Researchers showed that the decrease in intake of air leads to the decrease in fuel and decrease in engine torque for excess air ratio $\lambda = 1$ operation with 15% open throttle. It was discovered that a 1% increase in biomethane concentration in the intake air volume lowers engine output by 0.8%. It is required to advance the ignition angle by four crank angle degrees (CAD) to obtain optimum engine thermal efficiency.

Other works showed almost the same trend as presented in Sect. 2.3.1.1, even if the engine size and other operation conditions were different [25–34]. Thermal efficiency, output power, and NO_x emissions decreased with increase in CO_2 ratio, although HC (hydrocarbon), CO (carbon monoxide) and brake specific fuel consumption (BSFC) increased. The engine operation with methane-enriched biogas having 95% CH_4 shows almost the same level of performance as that with compressed natural gas (CNG) and much improved level of performance when 65% CH_4 and 32% CO_2 biogas was used [35]. The highest output power and efficiency were obtained for 50% biogas and 50% natural gas blend compared to biogas with propane or hydrogen blends [36, 37]. This means the fuel composed of 80% CH_4 and 20% CO_2 was the best for engine performance.

2.3.1.3   Other Research for Varying Equivalence Ratio

There have been many studies conducted on the effect of equivalence ratio on engine performance. The lean limit of methane-air mixture at atmospheric pressure and room temperature is about $\phi = 0.6$ ($\lambda = 1.67$). If in-cylinder temperature and pressure are higher, the lean limit can be extended a little more. Of course, to achieve stable operation of an engine, it is desirable that the excess air ratio does not exceed the lean limit. Biogas of 65% CH_4 and 35% CO_2 is used in an engine at $\lambda = 1.5$ [38]. The excess air ratio can be increased with hydrogen addition although much hydrogen may cause knocking, and the output power must be reduced to avoid knocking. For example, 10% H_2 addition enables the operation for excess air ratio of 1.5 [39]. The thermal efficiency operated with the mixture with 95% CH_4 (5% CO_2) with 20% H_2 at $\lambda = 1.7$ was higher than that with the mixture of stoichiometric 95% CH_4 (5% CO_2) [40].

2.3.2   Compression Ratio

Generally, thermal efficiency of a spark ignition engine increases with increase in compression ratio as indicated in equation of thermal efficiency of Otto cycle [41]. Figure 2.10 shows an example of experimental results of brake thermal efficiency in changing equivalence ratio and compression ratio from 9.3:1 to 15:1 [42]. The experimental conditions were engine speed at 1500 rpm, spark timing of MBT (minimum advance for best torque), full throttle, fuelled with biogas for a single cylinder engine. As the compression ratio increased, the peak thermal efficiency increased from 23 to 26.8% near stoichiometric condition. In particular, the thermal efficiency at compression ratio of 11:1 increased significantly compared to that of 9.3:1. The increase in compression ratio extended operating range to leaner condition at 0.64 for compression ratio 15:1 and at 0.77 for compression ratio 9.3:1. Higher compression ratio makes the mixture temperature increase and makes the residual gas smaller, so that lean limit of the flame propagation extends to leaner condition. This trend does not change regardless of engine size, etc. [31, 43–47]. Researchers also showed knocking threshold as the engine operation limit is related to the lean mixture, allowing to operate the engine at higher compression ratio [36].

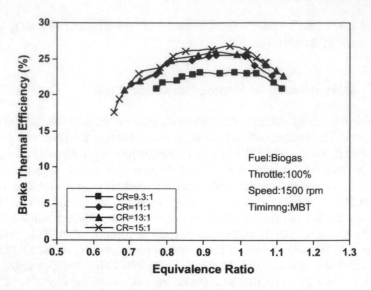

Fig. 2.10 Effect of equivalence ratio at full throttle on brake thermal efficiency in changing compression ratio [42]

2.3.3 Hydrogen Addition

When CO_2 ratio increases, the combustion becomes worse as described above. Then, hydrogen can be added to recover the weakness of combustion, especially at lean conditions [29, 33, 38, 40, 48–58]. Hydrogen has the drawbacks of producing much higher NO_x emissions due to higher combustion temperature, as well as a decrease in thermal efficiency due to large heat losses at high H_2 addition [59, 60] and/or increases in knock tendencies [61]. For example, when only 3% of hydrogen was added to the landfill gas composed of 53% CH_4, 42% CO_2 and 5% N_2, the indicated thermal efficiency increased 15% from 33 to 38% in the CFR (cooperative fuel research) engine at compression ratio 12, spark timing 25° before TDC, equivalence ratio 0.6, intake temperature 303 K, intake pressure 98 kPa and engine speed 600 rpm [62]. For another example, in the lean burn engine operation, as shown in Fig. 2.11, thermal efficiency and NO_x emissions are related with addition of H_2, as a function of excess air ratio at MBT spark timing [53, 54]. Thermal efficiency for 5% H_2 addition showed the maximum value for $\lambda = 1.0$. The thermal efficiency decreased with increase in H_2 addition greater than 10%. The thermal efficiency increased with the increase in excess air ratio except for the condition of $\lambda = 1.4$. The thermal efficiency in H_2 addition of 10% became larger than that in H_2 addition of 5% at $\lambda = 1.0$. The NO_x emissions showed the maximum value in $\lambda = 1.1$ and decreased with the increase in excess air ratio. At $\lambda = 1.4$ for H_2 addition of 10 and 20% the NO_x standard level was achieved.

As previously stated, the high CO_2 ratio causes sluggish combustion, which is the primary drawback of traditional AD biogas when utilized in SI engines. Researchers

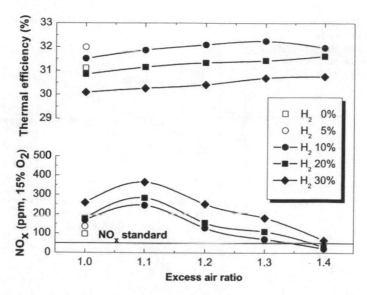

Fig. 2.11 Effect of excess air ratio on thermal efficiency and NO_x emissions for different concentrations of H_2 in biogas [53, 54]

also studied a two-phase anaerobic digestion (T-PAD) method, which generates H_2-enriched biogas [63]. The resulting T-PAD biogas has a typical composition of 10–25% H_2, 60–65% CH_4, and 15–30% CO_2. The combustion, performance, and emission characteristics of hydrogen-enriched T-PAD biogas was examined experimentally and computationally in a small-size SI engine, utilizing CH_4–H_2–CO_2 mixtures. They observed conditions when increasing the CO_2 ratio in the input biogas resulted in reduced engine power while increasing the H_2/CH_4 ratio resulted in increased engine power, particularly when the excess air ratio was more than 1.4. When excess air ratio was greater than 1.4, increasing the H_2/CH_4 ratio from 0 to 3/7 resulted in increased overall thermal efficiency, reduced level of brake-specific unburned hydrocarbons and reduced CO emissions. This allowed keeping brake-specific NO_x emissions at a very low level, which was lower than 2 g/(kWh) and this was due to more complete combustion. Increasing the H_2/CH_4 ratio from 1.2 to 1.4 had no influence on engine performance and emission characteristics, such as engine power, thermal efficiency, HC, and CO emissions, when the engine was operating at high loads, but it did substantially increase specific NO_x emissions owing to higher in-cylinder temperatures. T-PAD technology is still under development since its operating costs are greater due to the required sophisticated technical process. However, in the nearest future this technology is expected to reach the market that will allow using biogas blended with high H_2 concentrations.

Fig. 2.12 Changes in
thermal efficiency for
different spark plug types for
biogas blended with 10 and
30% H_2 [54]

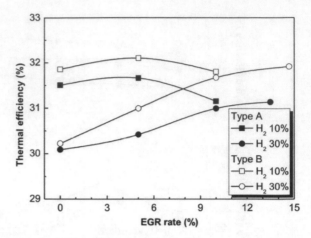

2.3.4   EGR

Exhaust gas recirculation (EGR) is very effective for reducing NO_x when combustion is activated owing to hydrogen addition, higher compression ratio, stoichiometric condition etc. [43, 44, 51]. However, EGR rate should be adjusted adequately to balance the promotion and suppression of the combustion. A new type of spark plug, with electrode gap positioned in the chamber, was used with hydrogen addition and EGR in biogas [54]. Figure 2.12 shows the effect of EGR rate on thermal efficiency at the spark timing of MBT. Here, the spark plugs of type A and B were the original one and newly designed one with electrode gap, positioned in the chamber. The peak thermal efficiency of 10% H_2 blends for type A increased 0.15% in the EGR case and 0.6% in the lean scenario compared to the initial state when the spark gap was anticipated, assuming the peak stayed fixed at a 5% EGR rate. The thermal efficiency with 30% H_2 rose as the EGR rate increased. For type B, the maximum permissible EGR rate was increased from 13.5 to 14.7%. When B type of spark plug was used, it was possible to achieve 0.8% increase in thermal efficiency. The authors claimed that this trend became significant at 30% H_2 addition, meaning that for high EGR rates and fuels with high H_2-blending ratios, larger spark gap projection into the chamber could be more effective. Comparable studies revealed similar outcomes utilizing hydrogen-blended low calorific gas when EGR and lean mixtures were used (40% natural gas and 60% nitrogen) [39].

2.3.5   Fuel Property

Landfill gas was used in a spark ignition engine and the performance was confirmed without any derating from the standard natural gas rating [64, 65]. An engine for operating heat pump was driven successfully with landfill gas in 70–90% rated speed range [66]. Because gasoline engine is widely used in the market, biogas was used instead of gasoline by changing fuel supply system from carburetor or fuel injector to other special carburetor or gas mixer of fuel and air [67]. When the engine was fueled with biogas instead of gasoline, 18% of reduction in brake power, 12% of reduction in brake thermal efficiency at wide open throttle (WOT) condition, as well as the decrease in CO and NO_x emissions were recorded [68]. To increase the engine performance, LPG was added 20% to the biogas [67]. An engine was operated for biogas, LPG and gasoline with WOT and stoichiometric condition [69]. It was shown the decrease in brake thermal efficiency, CO, HC, and NO_x emissions in biogas operation. Natural gas and biogas blend was tested and the control algorithm for stable operation was developed for varying fuel composition and output power [70]. The 5% syngas addition to landfill gas was found to not only significantly reduce CO, HC and NO_x emissions but also to improve brake efficiency of the engine [71]. Biogas (65% CH_4-35% CO_2) and ethanol were used for two different fuel replacements with biogas (20% and 50% by energy) [72]. They showed that 20% fuel replacement achieved larger IMEP (indicated mean effective pressure) values with lower NO_x emissions compared to ethanol single-fuel condition.

The effect of oxygen addition from 21 to 23% was investigated in a spark ignition biogas engine [73]. As oxygen ratio increased, peak brake thermal efficiency and brake power increases and the lean limit was extended. The exhaust emissions of CO and HC decreased although NO_x increased. However, NO_x remained within the regulation level.

2.3.6 Other Physical Parameters

The effect of preheating the inlet of biogas on operation by applying a waste-heat recovery system was investigated with 73% CH_4 and the result showed the engine performance was improved when the excess air ratio was relatively high, larger than 1.3 [74]. The reinforcement of turbulence in the cylinder was very effective to promote the combustion. The swirl flow generated with a masked valve was produced in the cylinder [75]. The modification of piston configuration also generated strong turbulence, as estimated with CFD (computational fluid dynamics) simulation [76]. The turbulence generated owing to the swirl improved combustion, led to the increase in power output and thermal efficiency although NO_x emissions increased. In a lean burn engine, fuel injection to manifold and twin spark plug led to higher thermal efficiency especially at low loads [77].

2.4   Knocking and Methane Number

2.4.1   *Visualization of Autoignition in the End-Gas Region and Knock Occurrence*

Abnormal combustion is mainly classified as spark knock and surface or localized spontaneous ignition. Spark knock is controllable by retarding the spark timing. Generally, when spark timing is advanced, thermal efficiency gradually increases. However, before the spark timing reaches MBT, knock sometimes occurs at high load. Once knocking occurred, the engine can be damaged due to pressure oscillation after autoignition of the end-gas in the cylinder. Such knock can occur not only in spark ignition engines but also in dual-fuel engines because of the autoignition of the end gas.

Figure 2.13 shows an example of time series images (A ~ I) of knock occurrence in the end-gas region and pressure history in a compression-expansion machine [78].

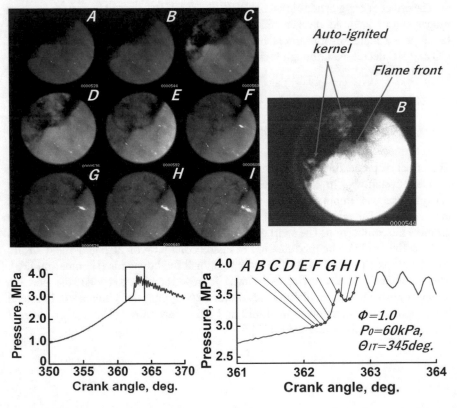

Fig. 2.13 Time series images of knock occurrence in the end-gas region (Top view; visualization area of 32 mm for bore of 78 mm) and the pressure history (same condition of Fig. 2 in [78])

The experimental conditions were engine speed of 600 rpm and the ignition timing θ_{ig} = 345 CAD with compression ratio of 9.0. The mixture was homogeneous n-C_4H_{10} + 6.5 O_2 + 26 Ar for equivalence ratio ϕ = 1.0. Argon was used instead of nitrogen to increase the unburned gas temperature and pressure. A window with diameter of ϕ32 mm was set at the head side of the end-gas region while ignition location was the opposite side, and the bore was 78 mm. The flame progressed from the right and bottom side to left and upper side of the bore. The symbols A ~ I correspond to those in magnified pressure history. The frame rate was 63,000 fps and the duration of flame images A ~ I was between 362.1 and 362.7 CAD. The image at timing B is magnified with enhanced image processing in the right-hand side of the figure. The occurrence of autoignition in the end-gas region was confirmed. Thereafter, pressure wave oscillations happened due to back and forth motion of the pressure wave, which are not seen on A ~ I images in Fig. 2.13. The autoignition in the end-gas region depends on the pressure and unburned gas temperature, type of fuel, equivalence ratio, and oxygen concentration.

On the other hand, although there are some other abnormal combustion regimes, pre-ignition is representative of surface or localized spontaneous ignition. Once pre-ignition occurs, the pressure in the cylinder rises rapidly and can cause autoignition in the end-gas region and the engine damage due to extremely strong pressure oscillations. In any case, the abnormal combustion should be avoided to ensure stable engine operation.

2.4.2 Estimation of Autoignition Timing

It is important to estimate the autoignition timing to predict knock occurrence in the end-gas region. Although the term of ignition delay is sometimes used for initial combustion duration such as 1% of MFB in spark ignition engines, the duration between spark ignition and occurrence of autoignition is defined as ignition delay time relevant to knocking. In engine cylinders, the pressure and temperature change with crank angle. Therefore, an empirical equation proposed by Livengood-Wu and called Livengood-Wu integral is often used to estimate the ignition delay using instantaneous pressure and temperature at time t [79].

$$\int_0^{\tau_{id}} (1/\tau)dt = 1 \qquad (2.6)$$

where τ_{id} denotes the ignition delay in the engine cylinder.

Basic studies in shock tubes, constant-volume vessels, and rapid-compression machines have been conducted to obtain autoignition behavior of fuel–air mixtures. Ignition delay data from these experiments have usually been expressed by equations of the form [41]:

Fig. 2.14 Ignition delay of
CH_4, CO_2 and air mixture
[80]

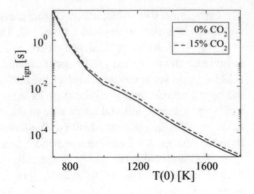

$$\tau = A_0 P^{-n} \exp\{E_a/(R_0 T)\} \tag{2.7}$$

where τ is ignition delay, A_0 and n are constants, and P, E_a, R_0 and T denote pressure, an apparent activation energy, the universal gas constant, and unburned gas temperature, respectively. Sometimes, the term on the effect of equivalence ratio, ϕ^{-m}, or other terms are added, where m is a constant.

Figure 2.14 shows the effect of unburned gas temperature on ignition delay of premixed methane-air diluted with 15% CO_2 at pressure of 3 MPa and equivalence ratio of 0.8 obtained through simulation [80]. In flame propagation with progress of combustion, the pressure and temperature of unburned gas increase gradually in the end-gas region. Knock occurs when the conditions satisfy Livengood-Wu integral shown in (2.6).

2.4.3 Methane Number

To avoid knocking is the most significant challenge for spark ignition engines. For gasoline engines, research octane number (RON) is often used to show the degree of resistance to knocking by employing two hydrocarbons. Normal heptane (n-C_7H_{16}) has octane number of 0 and iso-octane (i-C_8H_{18}, 2, 2, 4-trimethylpentene) has an octane number of 100. The octane number of a mixture of 10% n-heptane and 90% i-octane equals 90. Therefore, the octane number of a fuel is determined by comparing the knock resistance of the fuel based on the mix of n-heptane and i-octane according to the procedure of ASTM method D2699 [41].

Similarly, the methane number (MN) was developed to assess gaseous fuel resistance because the RON of gaseous fuel often exceeds 100. Leiker et al. came up with this parameter in 1972 [81]. A combination of methane (CH_4) and hydrogen (H_2) is used as the MN method's reference fuel. The knock characteristics of a test fuel matched by 100% CH_4 are assigned as MN of 100, while the knock characteristics of a test fuel matched by 100% H_2 are assigned as MN of 0. That is, 80% CH_4/20% H_2 combination has MN of 80. This is because methane has the strongest resistance to knocking of all the gaseous hydrocarbons, but hydrogen can be knocked very easily. Adding carbon dioxide to pure methane will, by definition, result in an MN greater than 100, which is calculated by adding the percentage of CO_2 added to 100. In other words, if the combination is 80% CH_4 and 20% CO_2, the MN is 120.

As described above, methane number (MN) is an indicator for knock limit of gaseous fuels. The knocking effect analysis was investigated for the fuel blend of biogas, natural gas, propane, and hydrogen [36, 37, 82–84]. Table 36 shows some examples of fuel properties such as lower heating value (LHV), low Wobbe index (LWI), energy density and methane number, % throttle and laminar burning velocity (S_L) for $\phi = 0.9$ at 1 atm [36]. They used a simulated biogas composed of 60% CH_4 and 40% CO_2. The blended fuels included biogas, methane, propane and hydrogen. Natural gas (NG) composition was 94.8% CH_4, 2.3% C_2H_6, 1.1% C_3H_8, 1.3% N_2 and 0.5% CO_2. The 50B50M mix was the same as pure biogas (80% CH_4 and 20% CO_2). As shown in Fig. 2.15, they achieved experimental findings for generating efficiency and maximum output power. The generating efficiency of the mixes was higher with greater output power. The MN of 50B50M was 120, and it had a greater energy density than biogas. The energy density of 57B38M5H and 54B36M10H blends was close to that of 50B50M fuel. These mix fuels, on the other hand, had a reduced knocking resistance. Because of the hydrogen presence, the burning was improved. To avoid knocking, the spark timing had to be retarded. The maximum pressure dropped, delaying the combustion process. Although the output power remained constant, this resulted in a decrease in efficiency. For 7.5 kW, 100B engine had an efficiency of 28.7%. As a result, 50B50M mix fuels were highly appealing.

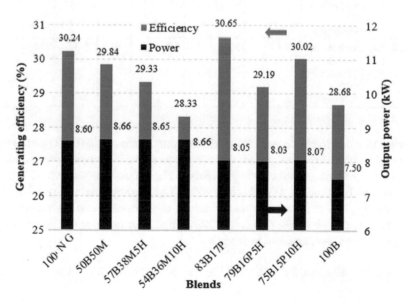

Fig. 2.15 Generating efficiency and maximum output power for several kinds of blended fuels [36]

Table 2.1 Properties of blended fuels [36]

Fuel designation	Fuel composition	Fuel properties				% Throttle (ETVO)	S_L (cm/s) $\phi =$ 0.9 at 1 atm
		LHV MJ/m³ fuel	LWI MJ/m³ fuel	Energy density MJ/m³ air	MN		
100NG	100% NG	346	45.1	3.6	87.2	61.7	35.0
100B	100% Biogas	20.4	21.0	3.4	140.0	75.0	23.6
50B50M	50% Biogas + 50% Methane	27.1	31.4	3.6	120.0	66.7	30.2
57B38M5H	57% Biogas + 38% Methane + 5% Hydrogen	25.0	29.0	3.6	105.3	63.3	30.3
54B36M10H	54% Biogas + 36% Methane + 10% Hydrogen	24.2	28.8	3.6	96.5	60.0	31.7
83B17P	83% Biogas + 17% Propane	32.0	31.4	3.6	65.8	58.3	32.8
79B16P5H	79% Biogas + 16% Propane + 5% Hydrogen	30.9	31.0	3.7	65.2	56.7	33.9
75B15P10H	75% Biogas + 15% Propane + 10% Hydrogen	30.7	30.7	3.8	63.8	55.0	35.0

Prediction and measurement of the critical compression ratio (CCR) and MN for blends of biogas were investigated because the increase in compression ratio leads to the increase in thermal efficiency [82]. Researchers experimented with 12 types of fuels such as 100 biogas (60% CH_4 and 40% CO_2), various mixtures of CH_4, CO_2, C_3H_8 and H_2, etc. with previous data obtained at Colorado State University. They proposed a relation between MN and CCR of engine for the biogas with other fuels such as methane, propane, and hydrogen.

$$MN = 0.1792(CCR)^3 - 6.9152(CCR)^2 + 96.236(CCR) - 398.16 \qquad (2.8)$$

As the value of MN increased, the critical compression ratio increased, and thermal efficiency would also increase. Such almost linear relation is shown in (2.8).

Methane number was also investigated for eight fuels of landfill gas (60% CH_4, 40% CO_2), digester gas (60% CH_4, 38% CO_2, and 2% N_2), coal gas, wood gas, and reformed natural gas [85]. They also showed linear relation between MN and compression ratio.

2.5 Pre-chamber Type for Lean Burn Engine

Pre-chamber ignition system is one of the effective technologies for lean burn engines, which was at first developed for automobile gasoline engine as CVCC (compound vortex controlled combustion) [86]. A review paper for pre-chamber ignition system was published [87]. There have been many studies on pre-chamber systems integrated in the cylinder head for natural gas engines [e.g., 88]. Multipoint ignition sources in main chamber were formed owing to flame jets issued from the orifices of the pre-chamber. In this experiment, the same rated brake power output, and the same NO_x emissions were obtained while CO and HC emissions were reduced by 15% and 8%, respectively. When the spark timing was advanced, further thermal efficiency increase was achieved under the Swiss regulation of CO and NO_x at that time.

Figure 2.16b shows schematic diagram of a pre-chamber equipped in a constant volume vessel of 300 mm diameter fueled with CH_4-air mixture at high ambient pressure [89]. Figure 2.16a shows an open chamber ignition system for comparison. The pre-chamber TJI (Turbulent Jet Ignition) system uses burned gas jet from the holes between the pre-chamber and the main chamber. The fuel–air condition in the main chamber was very lean, such as λ = 1.8–2.0. The combustion in the pre-chamber started due to the spark discharge. The lean condition requires stronger

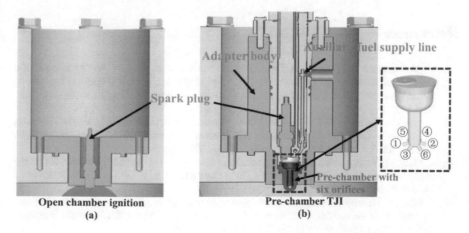

Open chamber ignition   **Pre-chamber TJI**
(a)   **(b)**

Fig. 2.16 Pre-chamber for large-size constant-volume vessel (diameter = 300 mm)[89]

spark energy, so that additional fuel was supplied to the pre-chamber by means of auxiliary supply line. The excess air ratio in the pre-chamber was richer ($\lambda - 1.0$) than that in the main chamber to ensure ignition. Thereafter, the burned gas jets passed through six holes and came out to the main combustion chamber because the pressure in the pre-chamber was higher than that in the main chamber. In gas engines used for power generation, the pressure in the cylinder is much higher because large bore engines are often used at super-charged condition. In these engines, stronger spark energy would be needed.

2.6   Simulation of Biogas Combustion in SI Engines

The simplest simulation method is zero-dimensional with Wiebe function for a biogas SI engine [90]. The effect of CO_2 ratio for biogas was investigated with a phenomenological model [34]. 3D-CFD simulations based on Reynolds-averaged Navier–Stokes (RANS) equations with many physical and chemical models using commercial well validated codes have recently gained popularity [91]. For example, after constructing and experimentally validating a KIVA4-based CFD simulation model of an SI engine coupled with CHEMKIN, this model was then applied to a biogas engine [92]. However, these models are not complete and still need to be validated under a real engine environment. Studying chemical processes to determine the fundamental combustion properties of ignition delay and laminar flame speed is crucial for simulation of processes in spark ignition engines. In addition, shock tubes, rapid compression machines (RCM), flow reactors, and jet-stirred reactors (JSR) have all been used to evaluate the ignition delay. In recent years, direct numerical simulation (DNS) and large eddy simulation (LES) of gas flow and combustion processes in engine cylinders have emerged (e.g., [93]).

References

1. M. Maizonnasse, J.S. Plante, D. Oh, C.B. Laflamme, Investigation of the degradation of a low-cost untreated biogas engine using preheated biogas with phase separation for electric power generation. Renew. Energy **55**, 501–513 (2013). https://doi.org/10.1016/j.biombioe. 2005.11.022
2. L. Martinez-Valencia, D. Camenzind, M. Wigmosta, M. Garcia-Perez, M. Wolcott, Biomass supply chain equipment for renewable fuels production: a review. Biomass Bioenergy **148**, 106054 (2021). https://doi.org/10.1016/j.biombioe.2021.106054
3. R. Maly, M. Vogel, Initiation and propagation of flame fronts in lean CH4-air mixtures by the three modes of the ignition spark. Symp. (Int.) Combust. **17**, 821–831 (1979). https://doi.org/10.1016/S0082-0784(79)80079-X
4. R. Maly, R. Herweg, Spark ignition combustion in four-stroke gasoline engines, in *Flow and Combustion in Reciprocating Engines*, 1–66, ed. by C. Arcoumanis, T. Kamimoto (Springer, Berlin, 2008). https://doi.org/10.1007/978-3-540-68901-0_1

5. B. Lewis, G. von Elbe, *Combustion, Flames and Explosions of Gases*, 3rd edn. (Academic Press, New York, 1987)

6. A.A. Konnov, A. Mohammad, V.R. Kishore, N.I. Kim, C. Prathap, S. Kumar, A comprehensive review of measurements and data analysis of laminar burning velocities for various fuel + air mixtures. Prog. Energy Combust. Sci. **68**, 197–267 (2018). https://doi.org/10.1016/j.pecs.2018.05.003

7. L. Pizzuti, C.A. Martins, P.T. Lacava, Laminar burning velocity and flammability limits in biogas: A literature review. Renew. Sustain. Energy Rev. **62**, 856–865 (2016). https://doi.org/10.1016/j.rser.2016.05.011

8. N. Hinton, R. Stone, Laminar burning velocity measurements of methane and carbon dioxide mixtures (biogas) over wide ranging temperatures and pressures. Fuel **116**, 743–750 (2014). https://doi.org/10.1016/j.fuel.2013.08.069

9. V.K. Yadav, A. Ray, M.R. Ravi, Experimental and computational investigation of the laminar burning velocity of hydrogen-enriched biogas. Fuel **235**, 810–821 (2019). https://doi.org/10.1016/j.fuel.2018.08.068

10. Z. Wei, H. Zhen, J. Fu, C. Leung, C. Cheung, Z. Huang, Experimental and numerical study on the laminar burning velocity of hydrogen enriched biogas mixture. Int. J. Hydrogen Energy **44**, 22240–22249 (2019). https://doi.org/10.1016/j.ijhydene.2019.06.097

11. H.O.B. Nonaka, F.M. Pereira, Experimental and numerical study of CO_2 effects on the laminar burning velocity of biogas. Fuel **182**, 382–390 (2016). https://doi.org/10.1016/j.fuel.2016.05.098

12. M.V. Petrova, F.A. Williams, A small detailed chemical-kinetic mechanism for hydrocarbon combustion. Combust. Flame **144**, 526–544 (2006). https://doi.org/10.1016/j.combustflame.2005.07.016

13. G.P. Smith, D.M. Golden, M. Frenklach, N.W. Moriarty, B. Eiteneer, M. Goldenberg, C.T. Bowman, R.K. Hanson, S. Song, W.C. Gardiner, Jr., V.V. Lissianski, Z. Qin, GRI-Mech 3.0. (2000). http://combustion.berkeley.edu/gri-mech/version30/text30.html. Accessed 3 Nov 2021

14. H. Wang, X. You, A.V. Joshi, S.G. Davis, A. Laskin, F. Egolfopoulos, C.K. Law, *USC Mech Version II. High-Temperature Combustion Reaction Model of $H_2/CO/C_1$-C_4 Compounds* (2007). http://ignis.usc.edu/USC_Mech_II.htm, Accessed 3 Nov 2021

15. A.A. Konnov, Detailed reaction mechanism for small hydrocarbons combustion, in *Release 0.5, Available as Electronic Supplementary Material* (2000); F.H.V. Coppens, J. De Ruyck, A.A. Konnov, The effects of composition on the burning velocity and nitric oxide formation in laminar premixed flames of $CH_4+H_2+O_2+N_2$. Combust. Flame **149**, 409–417 (2007). https://doi.org/10.1016/j.combustflame.2007.02.004

16. Y.L. Chan, M.M. Zhu, Z.Z. Zhang, P.F. Liu, D.K. Zhang, The effect of CO_2 dilution on the laminar burning velocity of premixed methane/air flames. Energy Procedia **75**, 3048–3053 (2015). https://doi.org/10.1016/j.egypro.2015.07.621

17. P. Zahedi, K. Yousefi, Effects of pressure and carbon dioxide, hydrogen and nitrogen concentration on laminar burning velocities and NO formation of methane–air mixtures. J. Mech. Sci. Technol. **28**, 377–386 (2014). https://doi.org/10.1007/s12206-013-0970-5

18. V.P. Karpov, A.N. Lipatnikov, P. Wolanski, Finding the Markstein number using the measurements of expanding spherical laminar flames. Combust. Flame **109**, 436–448 (1997). https://doi.org/10.1016/S0010-2180(96)00166-6

19. W. Zhang, M.E. Morsy, Z. Ling, J. Yang, Characterization of flame front wrinkling in a highly pressure-charged spark ignition engine. Exp. Therm. Fluid Sci. **132**, 110534 (2022). https://doi.org/10.1016/j.expthermflusci.2021.110534

20. F.A. Williams, *Combustion Theory*, 2nd ed. (The Benjamin/Cummings Publishing Co., Inc., 1985)

21. R. Borghi, Turbulent combustion modelling. Prog. Energy Combust. Sci. **14**, 245–292 (1988). https://doi.org/10.1016/0360-1285(88)90015-9

22. N. Peters, Laminar flamelet concepts in turbulent combustion. Proc. Combust. Inst. **21**, 1231–1250 (1988). https://doi.org/10.1016/S0082-0784(88)80355-2

23. C. Mounaïm-Rousselle, L. Landry, F. Halter, F. Foucher, Experimental characteristics of turbulent premixed flame in a boosted Spark-Ignition engine. Proc. Combust. Inst. **34**, 2941–2949 (2013). https://doi.org/10.1016/j.proci.2012.09.008
24. Y. Kim, N. Kawahara, K. Tsuboi, E. Tomita, Combustion characteristics and NO_x emissions of biogas fuels with various CO_2 contents in a micro co-generation spark-ignition engine. Appl. Energy **182**, 539–547 (2016). https://doi.org/10.1016/j.apenergy.2016.08.152
25. E. Sendzikiene, A. Rimkus, M. Melaika, V. Makareviciene, S. Pukalskas, Impact of biomethane gas on energy and emission characteristics of a spark ignition engine fuelled with a stoichiometric mixture at various ignition advance angles. Fuel **162**, 194–201 (2015). https://doi.org/10.1016/j.fuel.2015.09.019
26. R.J. Crookes, Comparative bio-fuel performance in internal combustion engines. Biomass Bioenergy **30**, 461–468 (2006). https://doi.org/10.1016/j.biombioe.2005.11.022
27. J. Huang, R.J. Crookes, Assessment of simulated biogas as a fuel for the spark ignition engine. Fuel **77**, 1793–1801 (1998). https://doi.org/10.1016/S0016-2361(98)00114-8
28. C. Jung, J. Park, S. Song, Performance and NO_x emissions of a biogas-fueled turbocharged internal combustion engine. Energy **86**, 186–195 (2015). https://doi.org/10.1016/j.energy.2015.03.122
29. M. Karagöz, S. Sarıdemir, E. Deniz, B. Çiftçi, The effect of the CO_2 ratio in biogas on the vibration and performance of a spark ignited engine. Fuel **214**, 634–639 (2018). https://doi.org/10.1016/j.fuel.2017.11.058
30. Y. Kurtgoz, M. Karagoz, E. Deniz, Biogas engine performance estimation using ANN. Eng. Sci. Technol. Int. J. **20**, 1563–1570 (2017). https://doi.org/10.1016/j.jestch.2017.12.010
31. E.C. Kwon, K. Song, M. Kim, Y. Shin, S. Choi, Performance of small spark ignition engine fueled with biogas at different compression ratio and various carbon dioxide dilution. Fuel **196**, 217–224 (2017). https://doi.org/10.1016/j.fuel.2017.01.105
32. C. Mokrane, B. Adouane, A. Benzaoui, Composition and stoichiometry effects of biogas as fuel in spark ignition engine. Int. J. Automotive Mech. Eng. **15**, 4036–5052 (2018). https://doi.org/10.15282/ijame.15.1.2018.11.0390
33. E. Porpatham, A. Ramesh, B. Nagalingam, Investigation on the effect of concentration of methane in biogas when used as a fuel for a spark ignition engine. Fuel **87**, 1651–1659 (2008). https://doi.org/10.1016/j.fuel.2007.08.014
34. S.Y. Jung, J. Park, Numerical prediction of effects of CO_2 or H_2 content on combustion characteristics and generation efficiency of biogas-fueled engine generator. Int. J. Hydrogen Energy **42**, 16991–16999 (2017). https://doi.org/10.1016/j.ijhydene.2017.05.220
35. R. Chandra, V.K. Vijay, P.M.V. Subbarao, T.K. Khura, Performance evaluation of a constant speed IC engine on CNG, methane enriched biogas and biogas. Appl. Energy **88**, 3969–3977 (2011). https://doi.org/10.1016/j.apenergy.2011.04.032
36. J.P. Gómez Montoya, A.A. Amell, D.B. Olsen, G.J. Amador Diaz, Strategies to improve the performance of a spark ignition engine using fuel blends of biogas with natural gas, propane and hydrogen. Int. J. Hydrogen Energy **43**, 21592–21602 (2018). https://doi.org/10.1016/j.ijhydene.2018.10.009
37. J.P. Gómez Montoya, A.A. Amell, D.B. Olsen, Operation of a spark ignition engine with high compression ratio using biogas blended with natural gas, propane, and hydrogen. Trans. ASME J. Eng. Gas Turbines Power **141**, 051006 (2019). https://doi.org/10.1115/1.4041755
38. W.C. Nadaleti, G. Przybyla, SI engine assessment using biogas, natural gas and syngas with different content of hydrogen for application in Brazilian rice industries: efficiency and pollutant emissions. Int. J. Hydrogen Energy **43**, 10141–10154 (2018). https://doi.org/10.1016/j.ijhydene.2018.04.073
39. S. Lee, C. Park, S. Park, C. Kim, Comparison of the effects of EGR and lean burn on an SI engine fueled by hydrogen-enriched low calorific gas. Int. J. Hydrogen Energy **39**, 1086–1095 (2014). https://doi.org/10.1016/j.ijhydene.2013.10.144
40. W.C. Nadaleti, G. Przybyla, B. Vieira, D. Leandro, G. Gadotti, M. Quadro, E. Kunde, L. Correa, R. Andreazza, A. Castro, Efficiency and pollutant emissions of an SI engine using biogas-hydrogen fuel blends: BIO60, BIO95, H20BIO60 and H20BIO95. Int. J. Hydrogen Energy **43**, 7190–7200 (2018). https://doi.org/10.1016/j.ijhydene.2018.02.133

41. J.B. Heywood, *Internal Combustion Engine Fundamentals* (2nd ed.) (McGraw Hill, 2018), p. 178, 492, 496
42. E. Porpatham, A. Ramesh, B. Nagalingam, Effect of compression ratio on the performance and combustion of a biogas fuelled spark ignition engine. Fuel **95**, 247–256 (2012). https://doi.org/10.1016/j.fuel.2011.10.059
43. R. Sadiq, R.C. Iyer, Experimental investigations on the influence of compression ratio and piston crown geometry on the performance of biogas fueled small spark ignition engine. Renew. Energy **146**, 997–1009 (2020). https://doi.org/10.1016/j.renene.2019.06.140
44. A.J. Chaudhari, S.K. Hotta, N. Sahoo, V. Kulkarni, Combined impact of compression ratio and re- circulated exhaust gas on the performance of a biogas fueled spark ignition engine. J. Renew. Sustain. Energy **11**, 013104 (2019). https://doi.org/10.1063/1.5045742
45. S.K. Gupta, M. Mittal, Assessing the influence of compression ratio on engine characteristics including operational limits of a biogas-fueled spark-ignition engine. Trans. ASME J. Eng. Gas Turbines Power **142**, 121008 (2020). https://doi.org/10.1115/1.4048564
46. S.O. Bade Shrestha, G.A. Karim, Predicting the effects of the presence of diluents with methane on spark ignition engine performance. Appl. Therm. Eng. **21**, 331–342 (2001). https://doi.org/10.1016/S1359-4311(00)00039-9
47. S.K. Hotta, N. Sahoo, K. Mohanty, P. Mahanta, A.J. Chaudhari, Effect of compression ratio on the performance and emission characteristics of a raw biogas fueled spark ignition engine, in *Advances in Energy Research*, vol. 2, ed. by S. Singh, V. Ramadesigan. Springer Proc. in Energy. (Springer, Singapore, 2020), pp. 701–713. https://doi.org/10.1007/978-981-15-2662-6_63
48. C. Jeong, T. Kim, K. Lee, S. Song, K.M. Chun, Generating efficiency and emissions of a spark-ignition gas engine generator fuelled with biogas–hydrogen blends. Int. J. Hydrogen Energy **34**, 9620–9627 (2009). https://doi.org/10.1016/j.ijhydene.2009.09.099
49. J. Park, S. Song, Predicting the performance and NO_x emissions of a turbocharged spark-ignition engine generator fueled with biogases and hydrogen addition under down-boosting condition. Int. J. Hydrogen Energy **39**, 8510–8524 (2014). https://doi.org/10.1016/j.ijhydene.2014.03.150
50. C.D. Rakopoulos, C.N. Michos, Generation of combustion irreversibilities in a spark ignition engine under biogas–hydrogen mixtures fueling. Int. J. Hydrogen Energy **34**, 4422–4437 (2009). https://doi.org/10.1016/j.ijhydene.2009.02.087
51. K. Lee, T. Kim, H. Cha, S. Song, K.M. Chun, Generating efficiency and NO_x emissions of a gas engine generator fueled with a biogas-hydrogen blend and using an exhaust gas recirculation system. Int. J. Hydrogen Energy **35**, 5723–5730 (2010). https://doi.org/10.1016/j.ijhydene.2010.03.076
52. Y. Qian, S. Sun, D. Ju, X. Shan, X. Lu, Review of the state-of-the-art of biogas combustion mechanisms and applications in internal combustion engines. Renew. Sustain. Energy Rev. **69**, 50–58 (2017). https://doi.org/10.1016/j.rser.2016.11.059
53. C. Park, S. Park, Y. Lee, C. Kim, S. Lee, Y. Moriyoshi, Performance and emission characteristics of a SI engine fueled by low calorific biogas blended with hydrogen. Int. J. Hydrogen Energy **36**, 10080–10088 (2011). https://doi.org/10.1016/j.ijhydene.2011.05.018
54. C. Park, S. Park, C. Kim, S. Lee, Effects of EGR on performance of engines with spark gap projection and fueled by biogas-hydrogen blends. Int. J. Hydrogen Energy **37**, 14640–14648 (2012). https://doi.org/10.1016/j.ijhydene.2012.07.080
55. X. Zhang, J. Xu, S. Zheng, X. Hou, J. Liu, The experimental study on cyclic variation in a spark ignited engine fueled with biogas and hydrogen blends. Int. J. Hydrogen Energy **38**, 11164–11168 (2013). https://doi.org/10.1016/j.ijhydene.2013.01.097
56. S.O. Bade Shrestha, G.A. Karim, Hydrogen as an additive to methane for spark ignition engine applications. Int. J. Hydrogen Energy **24**, 577–586 (1999). https://doi.org/10.1016/S0360-3199(98)00103-7
57. S.O. Bade Shrestha, G. Narayanan, Landfill gas with hydrogen addition—a fuel for SI engines. Fuel **87**, 3616–3626 (2008). https://doi.org/10.1016/j.fuel.2008.06.019

58. Y. Karagöz, Analysis of the impact of gasoline, biogas and biogas + hydrogen fuels on emissions and vehicle performance in the WLTC and NEDC. Int. J. Hydrogen Energy **44**, 31621–31632 (2019). https://doi.org/10.1016/j.ijhydene.2019.10.019

59. J. Demuynck, N. Raes, M. Zuliani, M.D. Paepe, R. Sierens, S. Verhelst, Local heat flux measurements in a hydrogen and methane spark ignition engine with a thermopile sensor. Int. J. Hydrogen Energy **34**, 9857–9868 (2009). https://doi.org/10.1016/j.ijhydene.2009.10.035

60. T. Shudo, S. Nabeltani, Analysis of degree of constant volume and cooling loss in hydrogen fueled SI engine. SAE Tech. Paper 2001–01–3561, (2001). https://doi.org/10.4271/2001-01-3561

61. E. Porpatham, A. Ramesh, B. Nagalingam, Effect of hydrogen addition on the performance of a biogas fuelled spark ignition engine. Int. J. Hydrogen Energy **32**, 2057–2065 (2007). https://doi.org/10.1016/j.ijhydene.2006.09.001

62. G. Narayanan, S.O.B. Shrestha, Hydrogen as a Combustion Enhancer to Landfill Gas Utilization in Spark Ignition Engines. SAE Tech. Paper 2008-01-1040 (2008). https://doi.org/10.4271/2008-01-1040

63. Y. Zhang, M. Zhu, Z. Zhang, Y.L. Chan, D. Zhang, Combustion and emission characteristics of simulated biogas from two-phase anaerobic digestion (T-PAD) in a spark ignition engine. Appl. Therm. Eng. **129**, 927–933 (2018). https://doi.org/10.1016/j.applthermaleng.2017.10.045

64. G.P. Mueller, Landfill gas application development of the caterpillar G3600 spark-ignited gas engine. Trans. ASME J. Eng. Gas Turbines Power **117**, 820–825 (1995). https://doi.org/10.1115/1.2815470

65. N.C. Macari, R.D. Richardson, Operation of a Caterpillar 3516 spark-ignited engine on low Btu fuel. Trans. ASME J. Eng. Gas Turbines Power **109**, 443–447 (1987). https://doi.org/10.1115/1.3240061

66. J. Wu, Y. Ma, Experimental study on performance of a biogas engine driven air source heat pump system powered by renewable landfill gas. Int. J. Refrig. **62**, 19–29 (2016). https://doi.org/10.1016/j.ijrefrig.2015.08.023

67. I.W. Surata, T.G.T. Nindhia, I.K.A. Atmika, D.N.K.P. Negara, I.W.E.P. Putra, Simple conversion method from gasoline to biogas fueled small engine to powered electric generator. Energy Procedia **52**, 626–632 (2014). https://doi.org/10.1016/j.egypro.2014.07.118

68. S.K. Hotta, N. Sahoo, K. Mohanty, Comparative assessment of a spark ignition engine fueled with gasoline and raw biogas. Renew. Energy **134**, 1307–1319 (2019). https://doi.org/10.1016/j.renene.2018.09.049

69. S. Simsek, S. Uslub, Investigation of the impacts of gasoline, biogas and LPG fuels on engine performance and exhaust emissions in different throttle positions on SI engine. Fuel **279**, 118528 (2020). https://doi.org/10.1016/j.fuel.2020.118528

70. Y. Yamasaki, M. Kanno, Y. Suzuki, S. Kaneko, Development of an engine control system using city gas and biogas fuel mixture. Appl. Energy **101**, 465–474 (2013). https://doi.org/10.1016/j.apenergy.2012.06.013

71. M.P. Kohn, J. Lee, M.L. Basinger, M.J. Castaldi, Performance of an internal combustion engine operating on landfill gas and the effect of syngas addition. Ind. Eng. Chem. Res. **50**, 3570–3579 (2011). https://doi.org/10.1021/ie101937s

72. R.B.R. da Costa, R.M. Valle, J.J. Hernández, A.C.T. Malaquias, C.J.R. Coronado, F.J.P. Pujatti, Experimental investigation on the potential of biogas/ethanol dual-fuel spark-ignition engine for power generation: combustion, performance and pollutant emission analysis. Appl. Energy **261**, 114438 (2020). https://doi.org/10.1016/j.apenergy.2019.114438

73. E. Porpatham, A. Ramesh, B. Nagalingam, Experimental studies on the effects of enhancing the concentration of oxygen in the inducted charge of a biogas fuelled spark ignition engine. Energy **142**, 303–312 (2018). https://doi.org/10.1016/j.energy.2017.10.025

74. T.H. Lee, S.R. Huang, C.H. Chen, The experimental study on biogas power generation enhanced by using waste heat to preheat inlet gases. Renew. Energy **50**, 342–347 (2013). https://doi.org/10.1016/j.renene.2012.06.032

75. E. Porpatham, A. Ramesh, B. Nagalingam, Effect of swirl on the performance and combustion of a biogas fuelled spark ignition engine. Energy Convers. Manage. **76**, 463–471 (2013). https://doi.org/10.1016/j.enconman.2013.07.071

76. J.P. Gómez Montoya, A.A. Amell, Effect of the turbulence intensity on knocking tendency in a SI engine with high compression ratio using biogas and blends with natural gas, propane and hydrogen. Int. J. Hydrogen Energy **44**, 18532–18544 (2019). https://doi.org/10.1016/j.ijhydene.2019.05.146

77. G.S. Jatana, M. Himabindu, H.S. Thakur, R.V. Ravikrishna, Strategies for high efficiency and stability in biogas-fuelled small engines. Exp. Therm. Fluid Sci. **54**, 189–195 (2014). https://doi.org/10.1016/j.expthermflusci.2013.12.008

78. N. Kawahara, E. Tomita, Y. Sakata, Auto-ignited kernels during knocking combustion in a spark- ignition engine. Proc. Combust. Inst. **31**, 2999–3006 (2007). https://doi.org/10.1016/j.proci.2006.07.210

79. C.J. Livengood, C.P. Wu, Correlation of auto ignition phenomena in internal combustion engines and rapid compression machines. Proc. 5th Int. Symp. Combust. 347–356 (1955). https://doi.org/10.1016/S0082-0784(55)80047-1

80. E.A. Tingas, H.G. Im, D.C. Kyritsis, D.A. Goussis, The use of CO_2 as an additive for ignition delay and pollutant control in CH_4/air autoignition. Fuel **211**, 898–905 (2018). https://doi.org/10.1016/j.fuel.2017.09.022

81. M. Leiker, K. Christoph, Evaluation of antiknock property of gaseous fuels by means of methane number and its practical application to gas engines. ASME paper 72-DGP-4 (1972)

82. J.P. Gómez Montoya, A.A. Amell, D.B. Olsen, Prediction and measurement of the critical compression ratio and methane number for blends of biogas with methane, propane and hydrogen. Fuel 186, 168–175 (2016). https://doi.org/10.1016/j.fuel.2016.08.064

83. J.P. Gómez Montoya, G.J. Amador Diaz, A.A. Amell, Effect of equivalence ratio on knocking tendency in spark ignition engines fueled with fuel blends of biogas, natural gas, propane and hydrogen. Int. J. Hydrogen Energy **43**, 23041–23049 (2018). https://doi.org/10.1016/j.ijhydene.2018.10.117

84. J.P. Gómez Montoya, D.B. Olsen, A.A. Amell, Engine operation just above and below the knocking threshold, using a blend of biogas and natural gas. Energy **153**, 719–725 (2018). https://doi.org/10.1016/j.energy.2018.04.079

85. M. Malenshek, D.B. Olsen, Methane number testing of alternative gaseous fuels. Fuel **88**, 650–656 (2009). https://doi.org/10.1016/j.fuel.2008.08.020

86. T. Date, S. Yagi, A. Ishizuya, I. Fujii, Research and development of the Honda CVCC Engine. SAE Tech. Paper 740605 (1974). https://doi.org/10.4271/740605

87. C.E. Castilla Alvarez, G.E. Couto, V.R. Roso, A.B. Thiriet, R.M. Valle, A review of prechamber ignition systems as lean combustion technology for SI engines. Appl. Therm. Eng. **128**, 107–120 (2018). https://doi.org/10.1016/j.applthermaleng.2017.08.118

88. R.P. Roethlisberger, D. Favrat, Comparison between direct and indirect (prechamber) spark ignition in the case of a cogeneration natural gas engine, part I: engine geometrical parameters. Appl. Therm. Eng. **22**, 1217–1229 (2002). https://doi.org/10.1016/S1359-4311(02)00040-6

89. D. Ju, Z. Huang, X. Li, T. Zhang, W. Cai, Comparison of open chamber and pre-chamber ignition of methane/air mixtures in a large bore constant volume chamber: effect of excess air ratio and pre-mixed pressure. Appl. Energy **260**, 114319 (2020). https://doi.org/10.1016/j.apenergy.2019.114319

90. M.M.N. de Faria, J.P. Vargas Machuca Bueno, S.M.M.E. Ayad, C.R.P. Belchior, Thermodynamic simulation model for predicting the performance of spark ignition engines using biogas as fuel. Energy Convers. Manage. **149**, 1096–1108 (2017). https://doi.org/10.1016/j.enconman.2017.06.045

91. L.M. Dzikiti, P. Mukumba, CFD contextual modelling of biogas combustion in internal combustion engine: a review. Int. J. Eng. Res. Technol. **9**, 730–741 (2020). https://www.ijert.org/research/cfd-contextual-modelling-of-biogas-combustion-in-internal-combustion-engine-a-review-IJERTV9IS080295.pdf

92. X. Kan, D. Zhou, W. Yang, X. Zhai, C.H. Wang, An investigation on utilization of biogas and syngas produced from biomass waste in premixed spark ignition engine. Appl. Energy **212**, 210–222 (2018). https://doi.org/10.1016/j.apenergy.2017.12.037

93. M. Ameen, S. Patel, J. Colmenares, S. Wu, J. Chen, T. Nguyen, S. Desai, *Direct Numerical Simulation (DNS) and High-Fidelity Large-Eddy Simulation (LES) for Improved Prediction of In-Cylinder Flow and Combustion Processes*. DOE Vehicle Technologies Office Annual Merit Review (2021). https://www.energy.gov/sites/default/files/2021-06/ace146_ameen_2021_o_5-15_240am_LR_TM.pdf. Accessed 15 Nov 2021

Chapter 3
Combustion and Exhaust Emissions of Biogas Dual-Fuel Engines

Abstract Biogas can be utilized in dual-fuel engines because of higher thermal efficiency. Biogas is supplied from intake port and liquid diesel fuel is injected in the cylinder directly. The combustion starts from the autoignition of a mixture of vaporized liquid fuel, further igniting biogas and air mixture. The initial combustion occurs at multi points, leading to certain and stable ignition followed by turbulent combustion. At first, visualization of the dual-fuel combustion with micro pilot injection is presented. The effects of liquid fuel injected, biogas flow rate, load, carbon dioxide (CO_2) ratio in biogas, exhaust gas recirculation (EGR), compression ratio, H_2 addition, pre-heating and other parameters are reviewed based on the literatures. Next, an example of the combustion achieving higher output and thermal efficiency with micro pilot dual-fuel combustion is described, as well as exhaust emissions. After the premixed mixture is autoignited in the end-gas region in latter half of the combustion, pressure oscillation does not occur in some conditions and transition to abnormal knocking combustion is avoided. The effect of CO_2 ratio on PREMIER combustion was investigated and it was found that with higher concentrations of CO_2 it was easier to keep control of PREMIER combustion.

Keywords Dual-fuel engine · Gas engine · Biogas · Combustion · Exhaust emissions · Autoignition · Knocking · PREMIER combustion

Abbreviations

ATDC	After TDC
BDF	Bio diesel fuel
BGES	Biogas energy share
BSFC	Brake specific fuel consumption
BTDC	Before TDC
BTE	Brake thermal efficiency
CA	Crank angle
CH_4	Methane

© The Author(s), under exclusive license to Springer Nature Switzerland AG 2022   43
E. Tomita et al., *Biogas Combustion Engines for Green Energy Generation*,
SpringerBriefs in Applied Sciences and Technology,
https://doi.org/10.1007/978-3-030-94538-1_3

CI   Compression ignition
CME   Coconut methyl ester
CN   Cetane number
CO   Carbon monoxide
CO_2 Carbon dioxide
COV   Coefficient of variation
DEE   Diethyl ether
DME   Dimethyl ether
EGR   Exhaust gas recirculation
HMN   Heptamethylnonane
IMEP   Indicated mean effective pressure
HC   Hydrocarbon
JME   Jatropha methyl ester
KI   Knock intensity
KME   Karanja methyl ester
LPG   Liquefied petroleum gas
NO   Nitrogen oxide
NO_x Oxides of nitrogen
PBD   Polanga biodiesel
PM   Particulate matter
PME   Palm methyl ester
PREMIER   Premixed mixture ignition in the end gas region
ROHR   Rate of heat release
SME   Soybean methyl ester
TDC   Top dead center
THC   Total hydrocarbons

3.1 Introduction

A dual-fuel engine uses two kinds of fuel such as gaseous and liquid fuels. Gaseous fuel is supplied from an intake manifold or sometimes injected directly into the cylinder. Natural gas, liquified petroleum gas (LPG), hydrogen, producer gas, syngas, biogas, etc. are used. The liquid fuel of diesel fuel, bio-diesel fuel (BDF) etc. are injected directly into the engine cylinder to initiate ignition. In this chapter fundamentals of biogas dual-fuel engine are described.

Dual-fuel engines have been developed for utilization of alternative gaseous fuels and review papers were published [1–3]. In particular, natural gas engine with diesel fuel is widely used for automobile engines, and stationary engines for electric power generation. Green fuel blends used in diesel engines are summarized from the viewpoint of energy savings and pollution reduction [4, 5]. The potential work of biogas as well as hydrogen on dual-fuel mode in diesel engines was listed in recent review articles [6, 7]. It is important to mention that the combustion and emission processes in dual-fuel combustion mode are very sensitive to the composition and quantity of the gaseous fuel.

Laminar burning velocity of biogas is lower than that of other gaseous fuels, such as natural gas or LPG. It also has limited flammability when used as an engine fuel, as shown in Chap. 2, which does not give a favor for engine performance, because biogas has a lower heating value (LHV) due to the high concentration of CO_2, which may be up to 50%. However, the biogas can allow reaching higher autoignition temperatures, making it resistant to knocking.

There are some advantages for dual-fuel engines compared to spark ignition (SI) engines. First, thermal efficiency of dual-fuel engines is higher because of higher compression ratio than that in SI engines. Second, initial combustion becomes stable because of higher ignition energy and multiple ignition locations. The ignition energy depends on the amount of liquid fuel. For example, when even 1 mg of diesel fuel per cycle, as a micro pilot fuel, is injected into the cylinder, the energy released from that micro pilot fuel is about 40 J (diesel fuel: ~ 43 MJ/kg), which is one thousand times larger than an ordinary spark energy of 40 mJ in an SI engine. This is useful to burn low calorific fuels such as biomass gas derived from pyrolysis of wood tips. Third, there are several ignition locations owing to diesel injector. This makes the combustion duration short, so that the initial combustion becomes stable. Furthermore, thermal efficiency is higher because it is easy to operate in lean combustion. A diesel engine is often used as a dual-fuel engine under super charged or turbo charged conditions, so that higher output power is obtained even under lean combustion operation.

High autoignition temperature of biogas precludes it from being used directly in compression ignition (CI) engines, necessitating the use of an ignition source. The dual-fuel operation method is seen to be the best option for utilizing biogas in CI engines. The biogas fuel is mixed with the input air and sent via an intake port. Combustion cannot be instigated by just compressing the biogas-air mixture because autoignition temperature of biogas is higher than temperatures at the end of compression. As a result, before the compression top dead center (TDC), a liquid fuel (usually diesel or biodiesel) is injected directly into the cylinder as an ignition source. There have been a number of research on the use of biogas in dual-fuel CI engines. In this chapter, engine performance, combustion, and exhaust emissions in dual-fuel engines are summarized. At first, various effects of engine parameters and operating conditions on dual-fuel combustion are presented based on the published papers. Next, an example of micro pilot dual-fuel engine for higher thermal efficiency with PREMIER (Premixed mixture ignition in the end gas region) combustion is presented.

3.2 Visualization of Dual-Fuel Combustion

A single cylinder with water-cooled test engine was prepared for visualization of the combustion. It had 96 mm bore and 108 mm stroke with compression ratio of

16:1. The intake pressure was 100 kPa and gaseous fuel was supplied to intake pipe of which location was about two meters upstream from the intake valve. A common rail system was used for injection of diesel fuel. The injection pressure was 40 MPa and the amount was 2 mg/cycle. The injection timing was 8° BTDC (before TDC). The injector had four holes with diameter of 0.1 mm. A simulated biomass gas that imitates wood pyrolysis gas was used for this experiment. The fuel composition was 22.3% H_2, 27.6% CO (carbon monoxide), 23.2% CO_2, 2.7% CH_4 (methane) and 24.2% N_2 by volume. The total equivalence ratio ϕ was 0.6 and separately calculated equivalence ratio of diesel fuel was 0.03. The engine speed was 1000 rpm. The combustion in the cylinder was visualized directly through a sapphire window installed in the elongated piston, from the bottom with a high-speed color camera every 8000 frames per second (fps). The combustion chamber was a pancake type. The visualization area was 62 mm in diameter.

Figure 3.1 shows time series of dual-fuel combustion visualized in an engine cylinder from a bottom view. The gaseous fuel was introduced from an intake port. Next, although sprays cannot be seen, a small amount of diesel fuel was injected into the cylinder in four directions. The spray penetrates with entraining surrounding gaseous fuel and air. The mixtures were formed in four regions and autoignited after some delay time (ignition delay). Premixed flames developed gradually from the autoignited areas. The autoignited flame kernel showed blue color in this case, because the first premixed combustion regions were not rich condition. In some cases, the flame kernel showed a little bit luminous because of the existence of some fuel-rich regions. There were luminous parts due to incompleteness of injection that were remained as particles of diesel fuel after a while. This was because the diesel fuel that landed on the optical window started to burn. Next, the flame develops in the premixed mixture of gaseous fuel and air from the autoignited locations. So-called wrinkled flame structure can be seen clearly, although this was a direct photograph, and small droplets of the diesel fuel burned brightly.

Even if the engine specifications and experimental conditions are different, general initial combustion phenomena are almost same, as shown in Fig. 3.1, in the case of micro pilot injection that the biogas energy share (BGES) is very large. When the energy share of diesel fuel increases, the combustion behavior changes mainly to diffusion combustion of the diesel spray.

3.3 Effects of Various Parameters on Engine Performance and Exhaust Emissions

3.3.1 Biogas Flow Rate and Load

The biogas was confirmed to be used in a diesel engine as dual-fuel mode instead of petroleum fuel [8, 9]. The ratio of diesel substitution to biogas depends on the experiments.

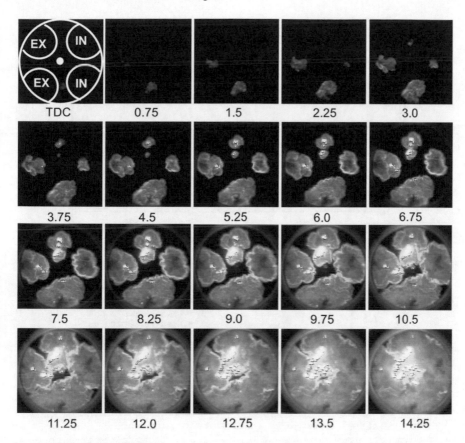

Fig. 3.1 Time series of flame images with a high-speed color camera of the combustion process. Intake pressure $P_{in} = 101$ kPa, injection timing $\theta_{inj} = 8°$BTDC, quantity of diesel fuel $= 2$ mg/cycle, injection pressure $P_{inj} = 40$ MPa, pyrolysis gas (22.3% H_2, 27.6% CO, 23.2% CO_2, 2.7% CH_4 and 24.2% N_2 in volume), total equivalence ratio $= 0.6$

For example, as shown in Fig. 3.2, the biogas induction increased with increase in BGES [10]. The composition of the biogas was 73% CH_4 and 17% CO_2. When the engine load increased from 0 to 100%, the BGES was decreased at the same biogas induction of 0.3, 0.6, 0.9 and 1.2 kg/h. In this experiment, BGES is not so large, so that diffusion combustion of the diesel spray is considered to play a main role. The engine performance parameters of brake specific fuel consumption (BSFC), brake thermal efficiency (BTE) and exhaust gas temperature are shown in Fig. 3.3 for diesel only and dual-fuel operation at constant injection timing of 23°BTDC [10]. The BSFC in dual-fuel mode was higher than that in diesel mode. This is due to the lower energy density of biogas, lower cylinder temperature, and slow burning owing to CO_2 in biogas. Because the BGES became larger at lower loads, the difference of BSFC was larger. The BTE in dual-fuel mode showed lower than that in diesel mode and decreased with the increase in BGES at all loads. At higher BGES, decrease in

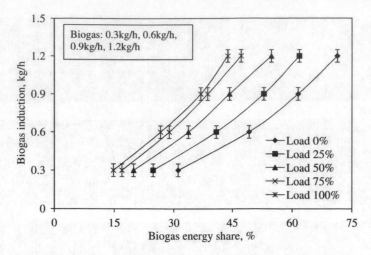

Fig. 3.2 Variation of biogas induction quantity with biogas energy share (BGES) [10]

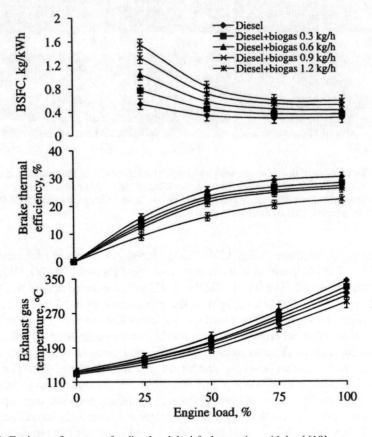

Fig. 3.3 Engine performances for diesel and dual-fuel operation with load [10]

oxygen concentration and lower gas temperature at injection timing due to higher specific heat with much CO_2 in the biogas caused longer ignition delay and slower diffusion combustion. In general, the exhaust gas temperature decreases with the increase in thermal efficiency in diesel mode. In this case, however, the exhaust gas temperature in dual-fuel mode was lower than that in diesel mode although the thermal efficiency was lower. The increase in BGES caused exhaust gas temperature to decrease because of the dilution of charge by presence of CO_2 in biogas.

Figure 3.4 shows specific exhaust gas emissions of CO, HC (hydrocarbon), NO (nitrogen oxide), CO_2, and smoke opacity with engine load [10]. The CO and HC emissions in the dual-fuel mode increased with the increase in BGES and decreased with the increase in engine load. This is due to incomplete combustion caused by dilution of charge with the presence of CO_2 in biogas. However, at high load, poor mixture formation of gaseous and liquid fuels may also be another reason for higher CO emission. The NO emission decreased with the increase in engine load and BGES. The NO emission depends on the burned gas temperature in the cylinder. The NO emission decreased with the increase in amount of CO_2, causing decrease in burned gas temperature. The CO_2 emission is an indication of complete combustion. The smoke opacity increased with the increase in engine load because of the combustion due to decrease in BGES and increase in diesel fuel. The smoke opacity decreased with the increase in BGES because of increase in gaseous fuel that does not include aromatics component in the fuel. Almost similar results for engine performance and exhaust emissions were obtained by other researchers [11–17].

On the other hand, the BTE in dual-fuel mode was comparable at full load or 80% load to diesel only mode [18–21]. At full load, the BTE was same or larger [22–24]. The BTE in dual-fuel mode was higher than that in diesel only mode at 80% load with increasing equivalence ratio from 0.35 to 0.7 and energy ratio of biogas to total energy of 0.6, by adjusting throttle valve [25].

In many experiments HC and CO in dual-fuel mode increased with increase in BGES [10–15, 17]. At middle loads, NO was larger than that in diesel mode at higher BGES [15].

In all research works with changing biogas flow rate and load, particulate matter (PM) or smoke in dual-fuel mode was reduced compared to those of diesel only operation [10, 11, 14, 15, 17, 20, 22–25]. A SEM (Scanning Electron Microscope) analysis showed that dual-fuel PMs were smaller and rounder than those of diesel [26].

Cycle-to-cycle fluctuations of maximum pressure and indicated mean effective pressure (IMEP) under the conditions of biogas proportion of 20 and 40% by mass were smaller than those in diesel mode, and oxides of nitrogen (NO_x) were also smaller [17].

The thermal efficiency in dual-fuel mode was strongly affected by biogas flow rate and methane ratio while the emissions of HC and CO had opposite trend. Therefore, optimum gas flow rate existed to give a better performance and lower emissions [10, 14, 22]. In biogas, the increases in HC, CO and NO_x were better controlled compared to that in producer gas [27].

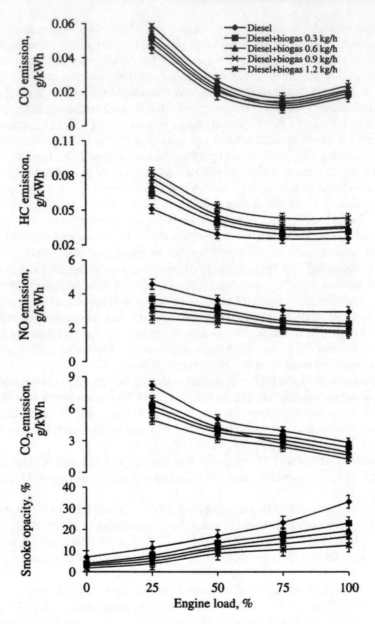

Fig. 3.4 Exhaust emissions for diesel and dual-fuel operation with load [10]

When biogas was supplied, the injection timing was advanced to achieve better engine performance because of slower burning velocity of biogas and air mixture [28]. Advancing injection timing showed that BTE, HC, CO and smoke were improved except the NO_x at full load. However, this NO_x value was lower than that in diesel mode [29].

Optimization of the engine performance and exhaust emissions was discussed using some parameters of load, compression ratio, injection timing [30], biogas flow rate, methane fraction of biogas, torque, charge temperature [31], and many other parameters of engine operation and performance [32]. The effect of *n*-butanol addition was also investigated, and positive results were obtained [33].

3.3.2 CO_2 Ratio in Biogas

Generally, as the ratio of CO_2 in biogas increases, NO_x decreases and HC and CO increase, due to slow combustion. However, the results of BTE are different among the published papers. The BTE was decreased with the increase in CO_2 ratio [34]. The BTE is often reduced at lower loads, although BTE was slightly reduced at higher load [28, 35, 36]. The BTE was lower at almost all conditions except some experimental conditions of 55% diesel fuel substitution, ~ 80% CH_4 in biogas and 2800 rpm [37]. The BTE in dual-fuel mode with changing CO_2 ratio was almost the same as that in diesel mode at high loads [38]. Knock resistance was higher with the increase in CO_2, and 30%CO_2 (70% CH_4) showed the best performance among dual-fuel modes [35]. The BTE with 70% CH_4 was higher than that with 60% CH_4 at biogas energy ratio of ~ 20%, 1500 rpm, and high load, which was larger than that in diesel only mode [22]. When CO_2 was supplied instead of natural gas with diesel fuel at the same load, BSFC was slightly lower at 30% CO_2 and almost the same at 40% CO_2 [39].

3.3.3 EGR

Generally, exhaust gas recirculation (EGR) reduces NO_x emissions due to the decrease in combustion gas temperature. The EGR ratio was changed up to 15% [40] and the results revealed that: At high load, the engine efficiency slightly decreased while HC and CO were almost the same; At low load, the engine efficiency slightly increased although HC and CO increased; The highest efficiencies were obtained at hot EGR case in both low and high loads; PM was increased at all loads. The BTE was recovered much in lower load and less in higher load although the BTE was lower in dual-fuel mode compared to diesel only mode [16, 20, 41], and CO and PM were increased [20]. The characteristics of dual-fuel engine were discussed at stoichiometric operation using high levels of EGR [42].

3.3.4 Liquid Fuel Injected

Ignition delay time τ for diesel engines is defined as the duration between the timing of fuel injection and occurrence of autoignition of the fuel and air mixture. The ignition delay depends on gas temperature, pressure, in-cylinder oxygen content, and the ratio of gaseous fuel induced to the intake port as presented in Sect. 2.4.2. The value of τ is estimated from Livengood-Wu Eq. (2.6) using instantaneous ignition delay derived from (2.7).

To evaluate ignition quality in a diesel engine, cetane number (CN) is often used. It is defined by blends of two pure hydrocarbon reference fuels: cetane (*n*-hexadecane, $C_{16}H_{34}$) with high ignition quality (CN = 100) and heptamethylnonane (HMN) with low ignition quality (CN = 15). Cetane number is given by

$$CN = \text{cetane (vol.\%)} + 0.15\,\text{HMN (vol.\%)} \tag{3.1}$$

A test fuel is operated in a CFR (cooperative fuel research) engine with variable compression ratio according to the procedure of ASTM Method D613. When the fuel presents the same ignition quality as the reference fuel of known CN, the CN of the test fuel is defined as the value of the known CN.

Figure 3.5 shows the effect of gas temperature on ignition delay time of four kinds of BDFs and diesel fuel at pressure P of 4 MPa and 2 MPa obtained from a constant-volume vessel [43]. Here, JME, CME, SME, PME denote jatropha methyl ester, coconut methyl ester, soybean methyl ester and palm methyl ester, respectively. These BDFs showed similar trends of ignition delay as that of diesel fuel. As the temperature in the vessel decreased, the ignition delay became longer due to slow chemical reaction. The ignition delay became shorter due to formation of more combustible gases at higher temperature. The CME and PME biofuels showed slightly shorter ignition delays than the others at $P = 2$ MPa. The negative temperature coefficient regimes between 750 and 850 K appeared at $P = 2$ MPa although they did not appear clearly at $P = 4$ MPa.

In some cases, various biodiesel or other high cetane number fuels are used instead of diesel fuel or are often mixed with diesel fuel. The liquid fuel properties affect ignition delay according to the cetane number. Generally, biodiesel fuels have disadvantages due to the higher density and viscosity that make fuel atomization worse, and also the higher boiling point that causes lower evaporation. However, smoke becomes less owing to oxygenated, low-sulfur, and non-aromatic fuel.

According to several literature sources, the performance of BDFs was slightly increased in BTE, decreased in CO and HC, and increased in NO_x [44]. These results were obtained with the combination of biodiesel and biogas [45, 46]. Karanja methyl ester (KME) was used instead of diesel fuel [47–51]. Waste cooking oil methyl ester [52] and 20% fish oil biodiesel [53] were tested and similar results were obtained. In 20% fish oil biodiesel with biogas, compatible engine performance was obtained and significantly lower HC and CO and slightly higher NO_x were recorded [53].

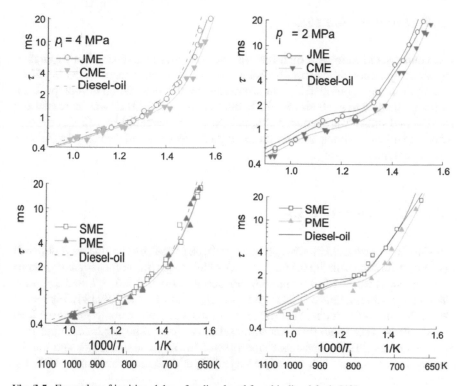

Fig. 3.5 Examples of ignition delays for diesel and four biodiesel fuels [43]

Rice bran oil methyl ester showed the best thermal efficiency, lowest CO and HC compared to other biodiesel of pongamia oil methyl ester and palm oil methyl ester, although the BTE was lower than that in diesel mode [54]. In 20% PBD (polanga biodiesel) with biogas, the BTE was lower, and HC and CO were larger with lower NO_x [55]. By optimization analysis, it was revealed that the engine performance in 20% PBD with 10% biogas was better than those of 20% PBD and 10% PBD with 10% ethanol.

When dimethyl ether (DME, CH_3OCH_3) was used as a liquid fuel, the shorter ignition delay was obtained because of high cetane number of DME [56]. When the ratio of biogas was increased, NO_x decreased and smoke was zero although HC and CO increased. Then, the IMEP decreased at late injection timing and increased at early timing. Different criteria were whether the spray of DME was injected into the piston bowl or toward the squish region.

Emulsified biodiesel may increase atomization owing to micro explosion of the fuel. The BTE with emulsified rice bran biodiesel with biogas was lower compared to that in diesel operation [57, 58]. The BTE increased for emulsified palm oil biodiesel with biogas compared to that of neat biodiesel operation [59]. It was also demonstrated that the BTE and exhaust gas temperature were increased with small amount of cerium oxide (CeO_2) nanoparticles added to the diesel fuel [60].

3.3.5 Compression Ratio

The compression ratio of a dual-fuel engine is higher than that of an SI engine. Higher compression ratio causes higher gas temperature and pressure near the TDC, leading to shorter ignition delay and promoting combustion. When the compression ratio was changed from about 16–19.5 in dual-fuel mode, the BTE was increased and HC, CO and smoke were reduced although NO_x was increased [40, 52, 57, 61, 62]. Advanced injection timing at some crank angle degrees showed the best performance in dual-fuel mode [63–65].

3.3.6 Hydrogen Addition

A review paper on the effect of biogas and hydrogen in dual-fuel engine was published [7]. Due to the better mixing of small amount of hydrogen, the combustion of gaseous fuels and air is expected to promote faster burning. Figures 3.6, 3.7, and 3.8 show examples of experimental results in changing H_2 addition up to 20% [24]. Figure 3.6 shows that the BTE in dual-fuel mode increased with the increase in engine load and showed higher value than that in diesel only mode at higher load, larger than 80%, while the BTE in dual-fuel mode showed lower value under low and middle engine load conditions. When the quantity of H_2 addition increased up to 15%, the

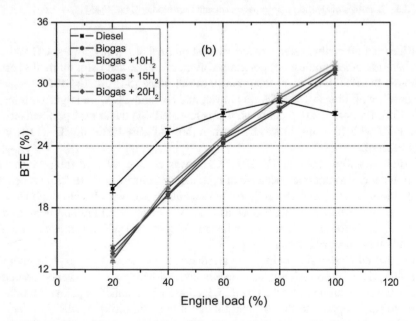

Fig. 3.6 Brake Thermal Efficiency (BTE) with engine load (biogas + H_2) [24]

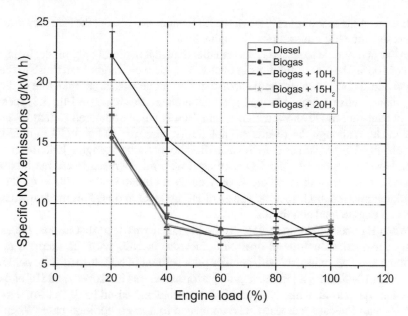

Fig. 3.7 Specific nitrogen oxide emissions with engine load (biogas + H$_2$) [24]

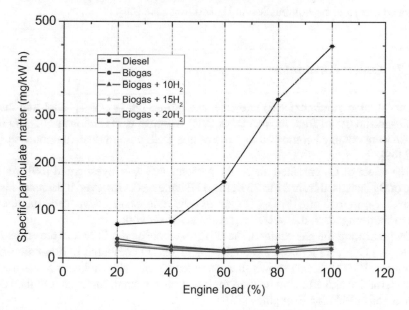

Fig. 3.8 Specific particulate matter emissions with engine load (biogas + H$_2$) [24]

BTE showed the maximum although the BTE value was almost the same. Figure 3.7 shows specific NO_x emissions with engine load.

In low and middle loads, NO_x was smaller than that in diesel only mode. In engine load of 100%, the NO_x emissions in dual-fuel mode were larger than those in diesel only mode. This is because the BTE in dual-fuel mode was larger than that in diesel only mode, causing higher burned gas temperature. As shown in Fig. 3.8, specific PM in dual-fuel mode showed significantly lower value compared to that in diesel only mode. Although the results are not shown here, the HC and CO in dual-fuel mode showed larger values than those in diesel only mode at engine load lower than 80%. However, both HC and CO decreased with the increase in engine load and showed smaller than those in diesel only mode at engine load of 100%. As the H_2 addition increased, both specific values of HC and CO in dual-fuel mode decreased under all engine load conditions.

When H_2 was added up to 15% in the biogas, it was seen that almost the same BTE as that in biofuel-diesel operation, increase in NO_x with the increase in H_2 addition, and PM reduction with smaller pilot amount at higher load were achieved [66]. With 10–20% H_2 addition, engine performance was improved, and HC and CO emissions decreased, while NO_x and PM emissions increased [67]. Dual-fuel mode with H_2 addition case increased BTE compared to non-H_2 addition case. When the injection timing of diesel fuel was retarded slightly, better engine performance was obtained due to faster combustion in H_2 addition case [68].

3.3.7 Other Parameters

When 4% diethyl ether (DEE) was injected into an intake port in dual-fuel mode of biogas and diesel fuel, the results of 2.3% increase in BTE, lower HC, CO and smoke were obtained compared to those of non-DEE case in dual-fuel mode at full load [49].

The effect of O_2 enriched air on the performance was investigated [69]. As the ratio of O_2 increased from 21 to 25%, the BTE increased. However, it decreased with the increase in load at 27% O_2. Higher oxygen ratio caused advanced combustion due to shorter ignition delay, causing much negative work.

By preheating the gas charged, the BTE was recovered a little and the emissions of HC and CO were reduced although NO_x increased compared to non-preheating operation [70–72]. It was shown that untreated biogas with preheating was a low-cost alternative and effective method for biogas use in small farms as it limited acid attacks and sulfidation in engines [73].

Supercharging and higher mixing conditions of biodiesel with biogas (60% CH_4 and 40% CO_2) showed higher BTE and lower CO and CH_4 emissions compared to naturally aspirated conditions of diesel fuel with biogas [74].

Higher nozzle opening pressure from 200 to 240 bar combined with higher compression ratio and advanced injection timing showed higher thermal efficiency [52, 75]. Post injection had no advantage in dual-fuel mode because the post-injected fuel was diffusively burned, while in pre-pilot injection before pilot injection, smoke emission was reduced because of improved charge homogeneity due to longer duration till main injection [15].

The main coating layer of 400 μm thickness with 8% yttria zirconia over a 100 μm thickness Ni–Al bond coating was processed on a piston as a thermal barrier [76]. Under dual-fuel mode, smoke was reduced, and NO_x was almost the same but slightly increased at higher loads while HC slightly decreased.

Simulations were performed using a double-Wiebe function [77]. Phenomenological two-zone model was used to predict dual-fuel combustion [78]. Recently, some commercial 3D-CFD (computational fluid dynamics) softwares based on RANS (Reynolds-averaged Navier-Stokes) equations are often used to predict the combustion and exhaust emissions in natural gas and diesel dual-fuel engines [79]. However, the simulation doesn't describe the combustion phenomena completely because of using many models as described in Sect. 2.6.

3.4 Higher Output Power and Thermal Efficiency with Micro Pilot Dual-Fuel Engine

This section explains the effect of CO_2 addition to CH_4 on engine performance characteristics and emissions such as NO_x, CO, HC. These include conditions at high loads of operations in stationary power generation. It is shown that the properties of end gas autoignition are critical under high load. The primary fuel used was biogas prepared by blending CH_4 and CO_2. Diesel fuel was injected as micro-pilot to ignite primary gaseous fuel. The pilot fuel injection timing was varied. The experimental tests were conducted in a supercharged dual-fuel engine using 200 kPa intake pressure with varied ratio of CO_2 to CH_4.

3.4.1 PREMIER Combustion

The behavior of end-gas autoignition without pressure oscillation at high load in dual-fuel engine was reported by Azimov et al. [80]. This mode of combustion is able to enhance engine performance with concurrent reduction of CO and HC emissions, which are normally increased in dual-fuel mode compared to those of a diesel mode. This mode of combustion was named "Premixed Mixture Ignition in the End-gas Region" (PREMIER). PREMIER combustion is different from knocking combustion when their end-gas autoignition properties are compared. In dual-fuel engines, pilot fuel autoignition causes air–fuel mixture ignition with proceeded flame propagation during which heat is released. The composition of gaseous fuel–air mixture which is heated in the end-gas region allows controlling the mode of combustion. For example, normal combustion occurs when, after the developed ignition kernel, the propagating flame with certain rate fully consumes the gaseous fuel–air mixture. If not all the fresh fuel–air mixture is consumed during the flame propagation, an autoignition in the end-gas region can occur. Such autoignition can be progressed as abnormal knocking combustion or as PREMIER combustion. The choice of combustion between two mentioned modes is driven by the pressure and unburned gas temperature histories in the end-gas region, and the combustion can be controlled by delaying the pilot fuel injection timing. It was experimentally demonstrated that the PREMIER combustion can boost the performance and efficiency of internal combustion engines [80–88]. Except for two studies [85, 88] where PREMIER combustion was observed in spark ignition engines, other investigations were mostly in dual-fuel gas engines. When syngas composed of different gases was used in PREMIER combustion mode, the IMEP and thermal efficiency were improved [81]. To suppress knocking combustion, different split injection strategies for pilot fuel were investigated to control the transition from normal combustion to PREMIER combustion and to suppress the transition from PREMIER combustion to knocking combustion [84]. PREMIER combustion intensity was used to measure the power produced by this combustion mode [86]. The prior research indicated that if the end-gas autoignition is achieved without severe pressure oscillations or without abnormal knocking combustion, this can help to improve engine thermal efficiency and potentially minimize the exhaust gas emissions such as CO and HC, though NO_x emissions somewhat increased due to the increased in-cylinder temperature.

3.4.2 Experimental Setup and Data Evaluation

The experiments were conducted using a four-stroke, single-cylinder, water-cooled dual-fuel engine with a displacement volume of 781 cm^3 and a compression ratio of 15.9:1, the engine had a diameter 96 mm and stroke 108 mm. During the tests, a piston having a shallow-dish top profile was used. The pilot fuel was injected using an injector driven by solenoid. The nozzle of the injector with three-holes of 0.011 mm in diameter each was specifically prepared for such experiments and controlled by a common rail system. This nozzle allowed injecting very small amount of diesel fuel. The equivalence ratio for micro-pilot diesel fuel, which was used as micro-pilot igniter, was 0.012, and the pilot fuel was injected at 40 MPa pressure at a flow rate of 1.6 mg/cycle [87]. The gaseous fuel was injected into an intake port and its supply into the engine cylinder was controlled using PLC to get correct gas compositions and fuel–air equivalence ratio. The biogas as a primary fuel was prepared by blending CH_4 with CO_2. All engine tests were conducted at 1000 rpm and 200 kPa intake pressure. The biogas-air charge equivalence ratio was maintained at 0.56 to achieve lean combustion. In the total equivalence ratio (biogas + diesel) the proportion of diesel fuel equivalence ratio was 2% only. Pilot fuel injection timing was advanced till knocking combustion occurred.

The gas composition is shown in Table 3.1. Gaseous fuels of CH_4 and CO_2 were supplied to the intake pipe more than one meter upstream of the intake valve to improve the homogeneity of the mixture. The energy value was adjusted from 2789

Table 3.1 Gas composition [87]

CO$_2$ ratio to CH$_4$ (vol.%)	Mass%			Heat value (J/cycle)
	Air	CH$_4$	CO$_2$	
0	96.85	3.14	0	2788.8
10	96.03	3.11	0.85	2761.9
20	95.33	3.09	1.57	2758.9
30	94.41	3.06	2.51	2756.5
40	93.61	3.03	3.34	2736.6
50	92.83	3.01	4.15	2720.1

to 2720 J/cycle by adding the CO$_2$ to CH$_4$ ratio from 0 to 50%, as indicated in Table 3.1. 4–20 kHz band-pass filter was used to process all in-cylinder pressure data to investigate pressure oscillations during knocking combustion. The occurrence of knocking cycles was regarded to be indicated by a knocking intensity (KI) with magnitude exceeding 0.1 MPa; this limit was chosen after subtracting the fluctuations produced by the noise. Any cycles which fell below this limit were classified as cycles with absence of any pressure oscillations (knock-free cycles) and were considered as normal or PREMIER combustion cycles [80]. The KI was then used to identify abnormal knocking combustion modes and distinguish them from normal or PREMIER combustion modes.

3.4.3 Results and Discussion

Data for the 20% CO$_2$ fueling conditions are provided in Fig. 3.9 as an illustration of the influence of pilot fuel injection timing on in-cylinder pressure and the rate of heat release (ROHR), where the maximum pressure in the cylinder rose as injection time was shifted to earlier positions [87]. In dual-fuel engines, three maxima in ROHR were observed, when PREMIER mode and knocking occurred. The first ROHR peak usually corresponds to the autoignition phase of pilot fuel. This isn't always obvious since when the ignition delay is short, as seen in Fig. 3.9, a lot of heat isn't released. The second peak, which is seen after the TDC usually corresponds to the gaseous fuel–air combustion by propagating flame. When end-gas autoignition occurs early, this is not always visible. Under PREMIER and knocking combustion circumstances, after the primary combustion phase due to flame propagation, the third peak can be visible due to the end-gas autoignition of the unburned mixture at a crank angle (CA) of roughly 10–15° ATDC (after TDC).

The third ROHR peak was seen in Fig. 3.9 for the $\theta_{inj} = -13°$ ATDC, $-14°$ ATDC, $-14.5°$ ATDC, and $-15°$ ATDC conditions, although there was no third peak at $\theta_{inj} = -12°$ ATDC (normal combustion). At $\theta_{inj} = -15°$ ATDC, several cycles were knocking, which showed high-frequency fluctuations in the pressure trace. The ROHR increased with the advancement of injection timing and PREMIER

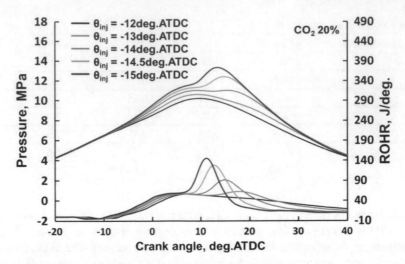

Fig. 3.9 Effect of injection timing on in-cylinder pressure and the rate of heat release at 20% CO_2 concentration in the mixture [87]

combustion occurred at $\theta_{inj} = -13°$ to $-14.5°$ ATDC. As the injection timing of pilot fuel was advanced, combustion mode shifted from normal combustion to knocking combustion through PREMIER combustion.

Figure 3.10 depicts the in-cylinder pressure and ROHR for pure CH_4 and 10, 20, 30, 40, and 50% CO_2 addition under ideal injection timing without knocking (PREMIER combustion). When 50% CO_2 was used, the in-cylinder pressure reached its maximum value close to TDC. When 10% CO_2 and pure CH_4 were used, the lowest in-cylinder pressure was detected. When the injection timing didn't change,

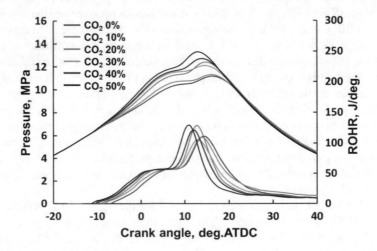

Fig. 3.10 Effect of CO_2 concentration on in-cylinder pressure and the rate of heat release [87]

raising the CO_2/CH_4 ratio tended to lower the peak of in-cylinder pressure [21, 36, 56], however, as seen in Fig. 3.9, increasing the injection timing resulted in a greater in-cylinder peak pressure. The addition of CO_2 to CH_4 might allow the pilot fuel injection time to be advanced across a larger range. The initial peak could not be observed clearly in the ROHR characteristics presented in Fig. 3.10; in particular, when CO_2 to CH_4 ratio was lower.

Figure 3.11 shows a conceptual illustration of PREMIER combustion that was observed in a dual-fuel engine [80]. In the first stage of this combustion mode, the pilot diesel fuel is injected, evaporated, and autoignited with gaseous fuel–air mixture prior to TDC. This autoignition initiates the flame development of the gaseous fuel–air mixture toward the cylinder wall. The second peak of ROHR often appears as shown in Fig. 3.11. Once the end-gas region is sufficiently heated and reaches the condition of autoignition, the second stage heat release begins. Thereafter, the third peak of ROHR is seen. The heat release is faster than that by only turbulent flame propagation. The start timing of PREMIER combustion is defined as the maximum timing of the second derivative of the ROHR.

Figure 3.12 depicts three combustion regimes (normal, PREMIER, and knocking combustion) shown with green, blue, and red hues, respectively, in varying CO_2 percentages at various pilot fuel injections. PREMIER combustion is registered when PREMIER cycles make up more than half of all cycles. When one or more knocking cycles appear, the mode is defined as knocking combustion. If PREMIER cycles appear at 49% of the experimental conditions, the combustion mode is still defined as normal combustion. PREMIER mode occurred with the advancement of injection of pilot fuel along with the increase of CO_2 to CH_4 ratio. Even at increasing CO_2 to CH_4 concentrations, the pilot injection range during PREMIER mode was kept

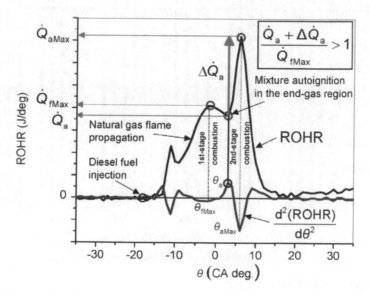

Fig. 3.11 Conceptual illustration of PREMIER combustion [80]

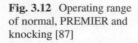

Fig. 3.12 Operating range
of normal, PREMIER and
knocking [87]

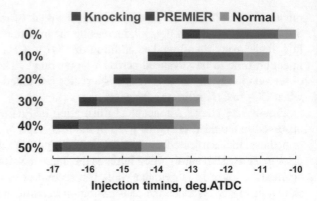

almost the same. The percentages of normal, PREMIER, and knocking combustion
cycles are shown in Fig. 3.13. Most of the cycles exhibited PREMIER mode at
$\theta_{inj} = -12.5°$ ATDC in the pure CH_4 example, however knocking happened more
than 20% of the time at $\theta_{inj} = -13.0°$ ATDC. When CO_2 was added, however,
and PREMIER combustion switched to knocking due to the CA's advanced time
(0.5°), the proportion of knocking was less than 10%. It was observed that during
the gradual increase in CO_2 concentration in the CO_2/CH_4 mixture, reaching 10, 30,
or 40%, knocking appeared in just one cycle. It was found to be easier to control the
PREMIER combustion in CO_2 addition case.

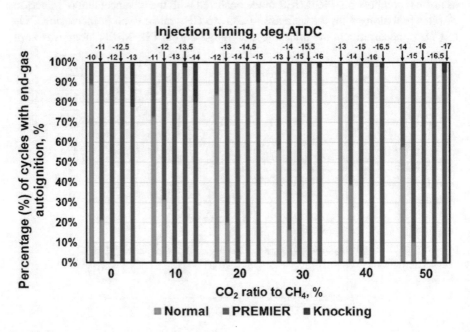

Fig. 3.13 End-gas autoignition (%) [87]

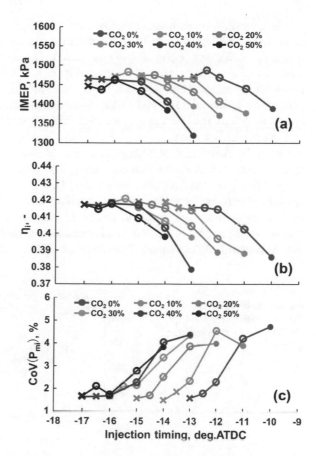

Fig. 3.14 Effect of CO_2 concentration on: (a) indicated mean effective pressure (IMEP); (b) indicated thermal efficiency; (c) coefficient of variation of the IMEP [87]

Engine performance characteristics such as IMEP, indicated thermal efficiency, and $COV_{(IMEP)}$ as a function of the injection timing of the CO_2–CH_4 mixture are shown in Figs. 3.14a–c, respectively. Here, the symbols of cross, open circle and closed circle denote knocking, PREMIER and normal combustion cases, respectively. Except in the 50% CO_2 situation, PREMIER combustion was achieved when injection timing was advanced and IMEP increased. Because combustion started earlier, a greater quantity of heat was produced in the advanced injection timing, resulting in increased in-cylinder pressure and IMEP. The highest IMEP was attained during PREMIER combustion for all conditions as a result of end-gas autoignition without pressure oscillation. Addition of CO_2 instigated the reduction in the IMEP although PREMIER mode was still present. For all situations, the advancement of injection timing boosted thermal efficiency, and PREMIER mode provided the highest thermal efficiency. When the CO_2 ratio was raised, other studies noticed a decrease in thermal efficiency [10, 21, 46], although the experimental conditions were different such as diesel fuel ratio that was much larger than that in this experiment, etc.

However, in the experiments described above, despite the reduction in heat release rate due to addition of more CO_2, the thermal efficiency during PREMIER mode remained unchanged or even slightly increased. The coefficient of variation (COV) indicates how stable an engine's functioning is in IMEP. $COV_{(IMEP)}$ was lowered under all situations when injection timing was advanced. Regardless of the CO_2 percentage in the mixture, PREMIER combustion increased operational stability. For all situations, the $COV_{(IMEP)}$ of the PREMIER operating mode was less than 5%.

Figures 3.15a–c depict NO_x, THC (total hydrocarbons), and CO emissions, respectively. Advancement of injection timing increased NO_x emissions in this investigation. It was shown that advancing injection timing increased NO_x emissions with a fixed overall equivalence ratio [3]. When injection timing is advanced, ignition occurs earlier in the process, burning more fuel and resulting in higher peak of in-cylinder temperatures, and as expected, such increase in temperature resulted in higher NO_x formation. In this experiment, knocking combustion produced the maximum NO_x emissions. The presence of CO_2 instigated reduction of NO_x

Fig. 3.15 Effect of CO_2 concentration on: (**a**) NO_x; (**b**) unburned hydrocarbon (THC); (**c**) carbon monoxide (CO) [87]

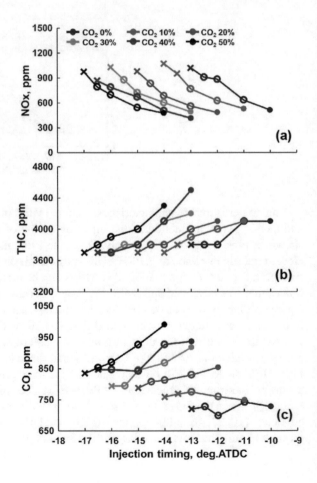

emissions during PREMIER combustion, especially with larger fractions of CO_2 in the CO_2–CH_4 mixtures. Because NO_x emissions are known to be temperature-dependent, increasing the amount of CO_2 in the mixture reduces NO_x emissions [46]. In this experiment, when higher CO_2/CH_4 ratios were used, a slight decrease in heat release instigated reduction of NO_x.

Because of the lower combustion temperature, THC and CO emissions rose when CO_2 fraction in the CO_2–CH_4 mixture increased in ref. [3, 46]. It was observed that PREMIER mode with favorable injection timing produced low level of THC and CO emissions, in particular, THC emissions at the PREMIER mode were nearly the same in all conditions, and CO emissions rose when CO_2/CH_4 ratio increased during PREMIER mode, as indicated in Fig. 3.15c.

The PM was estimated to be negligible because PM value measured with an opacimeter was not seen under the similar experimental conditions using natural gas [80]. Furthermore, 50% of CO_2/CH_4 ratio equates to only 2.7% CO_2 of the overall gas volume where the diesel fuel equivalence ratio was kept equal to 0.012. Since very little amount of diesel fuel was used, the PM level was nearly zero.

Figure 3.16 depicts the impact of ignition timing on ignition delay for various CO_2 to CH_4 ratios. A longer ignition delay was obtained by increasing the concentration of CO_2 or advancing injection timing. Because the in-cylinder temperature and pressure were lower at the time of fuel injection, with the advancement of injection timing, ignition was delayed. The temperature at ignition also reduced due to higher heat capacity as the CO_2 to CH_4 ratio increased. Consequently, it took longer for the mixture to ignite. Other researchers have shown similar findings [10, 36, 46, 56].

In-cylinder pressure at the start timing of autoignition in the end-gas region (θ_{ea}) is shown in Fig. 3.17 for various CO_2/CH_4 ratios. The average of 80 cycles is shown by each value. Normal, PREMIER, and knocking combustion are represented by closed circle, open circle, and cross symbols, respectively. The maximum pressure in PREMIER mode is connected by a dotted line as shown in Fig. 3.17. For PREMIER mode, the cylinder pressure at the end-gas autoignition was 10.6 MPa for the 0 and 10% CO_2 mixtures, 11.2–11.3 MPa for the 20 and 40% CO_2 mixtures, and 11.6 MPa for the 50% CO_2 mixtures. The in-cylinder pressure increased with the increase in CO_2 concentration in the mixture in PREMIER mode. Any abnormal knocking combustion was detected only above these pressure levels.

Fig. 3.16 Ignition delays during initial combustion [87]

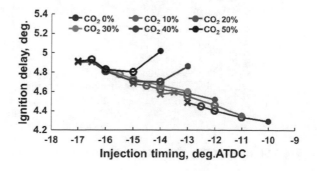

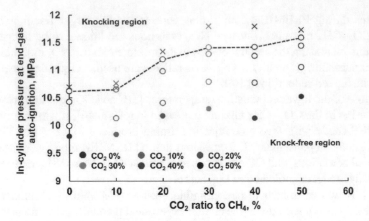

Fig. 3.17   Effect of CO_2–CH_4 ratio on in-cylinder pressure at the end-gas autoignition [87]

As the temperature of the unburned gas within the cylinder is critical in PREMIER combustion, the temperature was estimated using the in-cylinder pressure history and volume. There were assumed polytropic conditions in the cylinder during the period between the instance when the intake valve was closed and when diesel fuel was injected. Figure 3.18 displays the maximum temperature of the unburned gas at θ_{ea} for each CO_2 ratio in PREMIER mode as dotted line. Addition of CO_2 raised the temperature of the unburned gas at θ_{ea} and allowed avoiding knocking combustion. As the CO_2 ratio increased, so did the pressure and temperature in the end-gas region at the moment of autoignition. This is due to a decrease in the chemical reaction speed of unburned gas. As shown in Fig. 2.14, ignition delay with CO_2 addition is longer. It was also observed that even though the injection timing of pilot fuel was advanced, knocking did not occur. Therefore, it was determined that PREMIER combustion

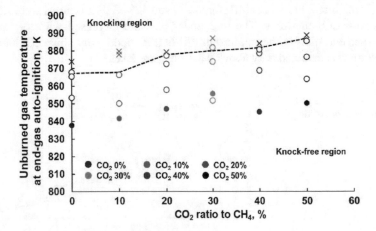

Fig. 3.18   Effect of CO_2–CH_4 ratio on unburned gas temperature at the end-gas autoignition [87]

could be attained by shifting the diesel injection timing to earlier crank angle positions and by concurrently adding more CO_2 into the primary gaseous fuel. Addition of high CO_2 concentrations helps avoiding knocking combustion even though the end gas region temperature may still be high.

As an outcome of the research, the following was concluded: (1) Even though the CO_2/CH_4 volume ratio increased, PREMIER mode was observed. With increasing CO_2 concentration, the maximum level of in-cylinder pressure increased under a favorable injection timing of pilot fuel. The addition of CO_2 instigated reduction of knocking, allowing to proceed with knock-free combustion cycles. (2) The advancement of pilot fuel injection timing instigated an increase in engine IMEP and thermal efficiency. As more CO_2 is added, it helped to improve PREMIER mode's thermal efficiency with unfortunate decrease in IMEP. (3) As the CO_2/CH_4 ratio increased, the maximum temperature of the unburned gas of PREMIER mode also increased. It was concluded that in order to achieve PREMIER combustion mode in dual-fuel gas engines, it is required attaining adequately high temperatures in the end-gas region with instigating autoignition and avoiding knocking.

References

1. G.A. Karim, A review of combustion processes in the dual fuel engine-the gas diesel engine. Progress Energy Combust. Sci. **6**, 277–285 (1980). https://doi.org/10.1016/0360-1285(80)900 19-2
2. G.A. Karim, Combustion in gas fueled compression Ignition engines of the dual fuel type. Trans. ASME J. Eng. Gas Turbines Power **125**, 827–836 (2003). https://doi.org/10.1115/1.158 1894
3. B.B. Sahoo, N. Sahoo, U.K. Saha, Effect of engine parameters and type of gaseous fuel on the performance of dual-fuel gas diesel engines—a critical review. Renew. Sustain. Energy Rev. **13**, 1151–1184 (2009). https://doi.org/10.1016/j.rser.2008.08.003
4. N.N. Mustafi, R.R. Raine, P.K. Bansal, The use of biogas in internal combustion engines—a review. ASME ICES2006 **1306**, 225–234 (2006). https://doi.org/10.1115/ICES2006-1306
5. J.K. Mwangi, W.J. Lee, Y.C. Chang, C.Y. Chen, L.C. Wang, An overview: Energy saving and pollution reduction by using green fuel blends in diesel engines. Appl. Energy **159**, 2140236 (2015). https://doi.org/10.1016/j.apenergy.2015.08.084
6. P. Rosha, A. Dhir, S.K. Mohapatra, Influence of gaseous fuel induction on the various engine characteristics of a dual fuel compression ignition engine: a review. Renew. Sustain. Energy Rev. **82**, 3333–3349 (2018). https://doi.org/10.1016/j.rser.2017.10.055
7. C. Deheri, S.K. Acharya, D.N. Thatoi, A.P. Mohanty, A review on performance of biogas and hydrogen on diesel engine in dual fuel mode. Fuel **260**, 116337 (2020). https://doi.org/10.1016/j.fuel.2019.116337
8. C. Jiang, T. Liu, J. Zhong, A study on compressed biogas and its application to the compression ignition dual-fuel engine. Biomass **20**, 53–59 (1989). https://doi.org/10.1016/0144-456 5(89)90020-6
9. A. Bilcan, O. Le Corre, A. Delebarre, Thermal efficiency and environmental performances of a biogas-diesel stationary engine. Environ. Technol. **24**, 1165–1173 (2003). https://doi.org/10.1080/09593330309385657
10. D. Barik, S. Murugan, Investigation on combustion performance and emission characteristics of a DI (direct injection) diesel engine fueled with biogas-diesel in dual fuel mode. Energy **72**, 760–771 (2014). https://doi.org/10.1016/j.energy.2014.05.106

11. D. Barik, M. Sivalingam, Performance and emission characteristics of a biogas fueled DI diesel engine. SAE Tech. Paper 2013-01-2507 (2013). https://doi.org/10.4271/2013-01-2507
12. D. Barik, A.K. Satapathy, S. Murugan, Combustion analysis of the diesel–biogas dual fuel direct injection diesel engine– the gas diesel engine. Int. J. Ambient Energy **38**, 259–266 (2015). https://doi.org/10.1080/01430750.2015.1086681
13. K. Cacua, L. Olmos-Villalba, B. Herrera, A. Gallego, Experimental evaluation of a diesel-biogas dual fuel engine operated on micro-trigeneration system for power, drying and cooling. Appl. Therm. Eng. **100**, 762–767 (2016). https://doi.org/10.1016/j.applthermaleng.2016.02.067
14. S.K. Mahla, V. Singla, S.S. Sandhu, A. Dhir, Studies on biogas-fuelled compression ignition engine under dual fuel mode. Environ. Sci. Pollut. Res. **25**, 9722–9729 (2018). https://doi.org/10.1007/s11356-018-1247-4
15. K.A. Rahman, A. Ramesh, Studies on the effects of methane fraction and injection strategies in a biogas diesel common rail dual fuel engine. Fuel **236**, 147–165 (2019). https://doi.org/10.1016/j.fuel.2018.08.091
16. M.S. Gaikwad, A.H. Kolekar, K.M. Jadhav, M.S. Yadav, P.R. Chitragar, Performance and emission characteristics of biomethane-diesel dual-fuelled CI engine in the presence of exhaust gas recirculation. Int. J. Ambient Energy (2020). https://doi.org/10.1080/01430750.2020.1722741
17. C. Jagadish, V. Gumtapure, Experimental studies on cyclic variations in a single cylinder diesel engine fuelled with raw biogas by dual mode of operation. Fuel **266**, 117062 (2020). https://doi.org/10.1016/j.fuel.2020.117062
18. P.M. Duc, K. Wattanavichien, Study on biogas premixed charge diesel dual fuelled engine. Energy Convers. Manage. **48**, 2286–2308 (2007). https://doi.org/10.1016/j.enconman.2007.03.020
19. C.C.M. Luijten, E. Kerkhof, Jatropha oil and biogas in a dual fuel CI engine for rural electrification. Energy Convers. Manage. **52**, 1426–1438 (2011). https://doi.org/10.1016/j.enconman.2010.10.005
20. V. Makareviciene, E. Sendzikiene, S. Pukalskas, A. Rimkus, R. Vegneris, Performance and emission characteristics of biogas used in diesel engine operation. Energy Convers. Manage. **75**, 224–233 (2013). https://doi.org/10.1016/j.enconman.2013.06.012
21. M. Feroskhan, S. Ismail, Investigation of the effects of biogas composition on the performance of a biogas-diesel dual fuel CI engine. Biofuels **7**, 593–601 (2016). https://doi.org/10.1080/17597269.2016.1168025
22. H. Ambarita, Performance and emission characteristics of a small diesel engine run in dual-fuel (diesel-biogas) mode. Case Stud. Therm. Eng. **10**, 179–191 (2017). https://doi.org/10.1016/j.csite.2017.06.003
23. H. Ambarita, Effect of engine load and biogas flow rate to the performance of a compression ignition engine run in dual-fuel (diesel-biogas) mode. IOP Conf. Ser.: Mater. Sci. Eng. **309**, 012006 (2018). https://iopscience.iop.org/article/10.1088/1757-899X/309/1/012006
24. R. Bouguessa, L. Tarabet, K. Loubar, T. Belmrabet, M. Tazerout, Experimental investigation on biogas enrichment with hydrogen for improving the combustion in diesel engine operating under dual fuel mode. Int. J. Hydrogen Energy **45**, 9052–9063 (2020). https://doi.org/10.1016/j.ijhydene.2020.01.003
25. F.Z. Aklouche, K. Loubar, A. Bentebbiche, S. Awad, M. Tazerout, Experimental investigation of the equivalence ratio influence on combustion, performance and exhaust emissions of a dual fuel diesel engine operating on synthetic biogas fuel. Energy Convers. Manage. **152**, 291–299 (2017). https://doi.org/10.1016/j.enconman.2017.09.050
26. N.N. Mustafi, R.R. Raine, A study of the emissions of a dual fuel engine operating with alternative gaseous fuels. SAE Tech. Paper 2008-01-1394 (2008). https://doi.org/10.4271/2008-01-1394
27. C. Nayak, S.S. Sahoo, L.N. Rout, Emission analysis of a dual fuel diesel engine fuelled with different gaseous fuels generated from waste biomass. Int. J. Ambient Energy **42**, 570–575 (2021). https://doi.org/10.1080/01430750.2018.1562971

28. S. Verma, L.M. Das, S.S. Bhatti, S.C. Kaushik, A comparative exergetic performance and emission analysis of pilot diesel dual-fuel engine with biogas, CNG and hydrogen as main fuels. Energy Convers. Manage. **151**, 764–777 (2017). https://doi.org/10.1016/j.enconman.2017.09.035

29. D. Barik, S. Murugan, Experimental investigation on the behavior of a DI diesel engine fueled with raw biogas-diesel dual fuel at different injection timing. J. Energy Inst. **89**, 373–388 (2016). https://doi.org/10.1016/j.joei.2015.03.002

30. S. Das, D. Kashyap, B.J. Bora, P. Kalita, V. Kulkarni, Thermo-economic optimization of a biogas-diesel dual fuel engine as remote power generating unit using response surface methodology. Therm. Sci. Eng. Progr. **24**, 100935 (2021). https://doi.org/10.1016/j.tsep.2021.100935

31. M. Feroskhan, S. Ismail, Evaluating the effect of intake parameters on the performance of a biogas–diesel dual-fuel engine using the Taguchi method. Biofuels **11**, 441–449 (2020). https://doi.org/10.1080/17597269.2017.1370885

32. A. Sharma, N.A. Ansari, A. Pal, Y. Singh, S. Lalhriatpuia, Effect of biogas on the performance and emissions of diesel engine fuelled with biodiesel-ethanol blends through response surface methodology approach. Renew. Energy **14**, 657–668 (2019). https://doi.org/10.1016/j.renene.2019.04.031

33. S.K. Mahla, S.M. Safieddin Ardebili, H. Sharma, A. Dhir, G. Goga, H. Solmaz, Determination and utilization of optimal diesel/n-butanol/biogas derivation for small utility dual fuel diesel engine. Fuel **289**, 119913 (2021). https://doi.org/10.1016/j.fuel.2020.119913

34. S.S. Kalsi, K.A. Subramanian, Effect of simulated biogas on performance, combustion and emissions characteristics of a bio-diesel fueled diesel engine. Renew. Energy **106**, 78–90 (2017). https://doi.org/10.1016/j.renene.2017.01.006

35. M.S. Lounici, K. Loubar, M. Tazerout, M. Balistrou, L. Tarabet, Experimental investigation on the performance and exhaust emission of biogas-diesel dual-fuel combustion in a CI engine. SAE Tech. Paper 2014-01-2689 (2014). https://doi.org/10.4271/2014-01-2689

36. S. Verma, L.M. Das, S.C. Kaushik, Effects of varying composition of biogas on performance and emission characteristics of compression ignition engine using exergy analysis. Energy Convers. Manage. **138**, 346–359 (2017). https://doi.org/10.1016/j.enconman.2017.01.066

37. A. Henham, M.K. Makkar, Combustion of simulated biogas in a dual-fuel diesel engine. Energy Convers. Manage. **39**, 2001–2009 (1998). https://doi.org/10.1016/S0196-8904(98)00071-5

38. N.N. Mustafi, R.R. Raine, S. Verhelst, Combustion and emissions characteristics of a dual fuel engine operated on alternative gaseous fuels. Fuel **109**, 669–678 (2013). https://doi.org/10.1016/j.fuel.2013.03.007

39. S. Bari, Effect of carbon dioxide on the performance of biogas/diesel dual-fuel engine. Renew. Energy **9**, 1007–1010 (1996). https://doi.org/10.1016/0960-1481(96)88450-3

40. S. Verma, L.M. Das, S.C. Kaushik, S.S. Bhatti, The effects of compression ratio and EGR on the performance and emission characteristics of diesel-biogas dual fuel engine. Appl. Therm. Eng. **150**, 1090–1103 (2019). https://doi.org/10.1016/j.applthermaleng.2019.01.080

41. M.S. Gaikwad, K.M. Jadhav, A.H. Kolekar, P.R. Chitragar, Combustion characteristics of biomethane–diesel dual-fueled CI engine with exhaust gas recirculation. Biofuels **12**, 369–379 (2021). https://doi.org/10.1080/17597269.2018.1479136

42. F. Königsson, P. Stalhammar, H-E. Angstrom, Characterization and potential of dual fuel combustion in a modern diesel engine. SAE Tech. Paper 2011-01-2223 (2011). https://doi.org/10.4271/2011-01-2223

43. C.W. Oo, M. Shioji, S. Nakao, N.N. Dung, I. Reksowardojo, S.A. Roces, N.P. Dugos, Ignition and combustion characteristics of various biodiesel fuels (BDFs). Fuel **158**, 279–287 (2015). https://doi.org/10.1016/j.fuel.2015.05.049

44. J.J. Hernández, M. Lapuerta, J. Barba, Effect of partial replacement of diesel or biodiesel with gas from biomass gasification in a diesel engine. Energy **89**, 148–157 (2015). https://doi.org/10.1016/j.energy.2015.07.050

45. S.H. Yoon, C.S. Lee, Effect of biofuels combustion on the nanoparticle and emission characteristics of a common-rail DI diesel engine. Fuel **90**, 3071–3077 (2011). https://doi.org/10.1016/j.fuel.2011.05.007

46. S.H. Yoon, C.S. Lee, Experimental investigation on the combustion and exhaust emission characteristics of biogas–biodiesel dual-fuel combustion in a CI engine. Fuel Proces. Technol. **92**, 992–1000 (2011). https://doi.org/10.1016/j.fuproc.2010.12.021
47. D. Barik, M. Sivalingam, Investigation on performance and exhaust emissions characteristics of a DI diesel engine fueled with Karanja methyl ester and biogas in dual fuel mode. SAE Tech. Paper 2014-01-1311 (2014). https://doi.org/10.4271/2014-01-1311
48. D. Barik, S. Murugan, Simultaneous reduction of NO_x and smoke in a dual fuel DI diesel engine. Energy Convers. Manage. **84**, 217–226 (2014). https://doi.org/10.1016/j.enconman.2014.04.042
49. D. Barik, S. Murugan, Effects of diethyl ether (DEE) injection on combustion performance and emission characteristics of Karanja methyl ester (KME)–biogas fueled dual fuel diesel engine. Fuel **164**, 286–296 (2016). https://doi.org/10.1016/j.fuel.2015.09.094
50. D. Barik, S. Murugan, N.M. Sivaram, E. Baburaj, P. Shanmuga Sundaram, Experimental investigation on the behavior of a direct injection diesel engine fueled with Karanja methyl ester-biogas dual fuel at different injection timings. Energy **118**, 127–138 (2017). https://doi.org/10.1016/j.energy.2016.12.025
51. D. Barik, A. Kumar, S. Murugan, Effect of compression ratio on combustion performance and emission characteristic of a direct injection diesel engine fueled with upgraded biogas-Karanja methyl ester-diethyl ether port injection. Energy Fuels **32**, 5081–5089 (2018). https://doi.org/10.1021/acs.energyfuels.7b01977
52. N. Khayum, S. Anbarasu, S. Murugan, Optimization of fuel injection parameters and compression ratio of a biogas fueled diesel engine using methyl esters of waste cooking oil as a pilot fuel. Energy **221**, 119865 (2021). https://doi.org/10.1016/j.energy.2021.119865
53. D.K. Ramesha, A.S. Bangari, C.P. Rathod, R.S. Chaitanya, Combustion, performance and emissions characteristics of a biogas fuelled diesel engine with fish biodiesel as pilot fuel. Biofuels **6**, 9–19 (2015). https://doi.org/10.1080/17597269.2015.1036960
54. B.J. Bora, U.K. Saha, Comparative assessment of a biogas run dual fuel diesel engine with rice bran oil methyl ester, pongamia oil methyl ester and palm oil methyl ester as pilot fuels. Renew. Energy **81**, 490–498 (2015). https://doi.org/10.1016/j.renene.2015.03.019
55. A. Sharma, Y. Singh, N.A. Ansari, A. Pal, S. Lalhriatpuia, Experimental investigation of the behaviour of a DI diesel engine fuelled with biodiesel/diesel blends having effect of raw biogas at different operating responses. Fuel **279**, 118460 (2020). https://doi.org/10.1016/j.fuel.2020.118460
56. S.H. Park, S.H. Yoon, J. Cha, C.S. Lee, Mixing effects of biogas and dimethyl ether (DME) on combustion and emission characteristics of DME fueled high-speed diesel engine. Energy **66**, 413–422 (2014). https://doi.org/10.1016/j.energy.2014.02.007
57. B.J. Bora, U.K. Saha, Improving the performance of a biogas powered dual fuel diesel engine using emulsified rice bran biodiesel as pilot fuel through adjustment of compression ratio and injection timing. Trans. ASME J. Eng. Gas Turbines Power **137**, 091505 (2015). https://doi.org/10.1115/1.4029708
58. B.J. Bora, U.K. Saha, Estimating the theoretical performance limits of a biogas powered dual fuel diesel engine using emulsified rice bran biodiesel as pilot fuel. Trans. ASME J. Energy Resour. Technol. **138**, 021801 (2016). https://doi.org/10.1115/1.4031836
59. B.K. Debnath, B.J. Bora, N. Sahoo, J.K. Saha, Influence of emulsified palm biodiesel as pilot fuel in a biogas run dual fuel diesel engine. ASCE J. Energy Eng. **140** (2014). https://doi.org/10.1061/(ASCE)EY.1943-7897.0000163
60. M. Feroskhan, S. Ismail, A. Kumar, V. Kumar, S.K. Aftab, Investigation of the effects of biogas flow rate and cerium oxide addition on the performance of a dual fuel CI engine. Biofuels **8**, 197–205 (2017). https://doi.org/10.1080/17597269.2016.1215072
61. B.J. Bora, U.K. Saha, S. Chatterjee, V. Veer, Effect of compression ratio on performance, combustion and emission characteristics of a dual fuel diesel engine run on raw biogas. Energy Convers. Manage. **87**, 1000–1009 (2014). https://doi.org/10.1016/j.enconman.2014.07.080
62. H.N. Singh, A. Layek, Experimental exploration of the impact of compression ratio on the characteristics of a biogas fueled dual fuel compression ignition engine. Int. J. Renew. Energy Res. **8**, 2075–2084 (2018). https://www.ijrer.org/ijrer/index.php/ijrer/article/view/8377

63. B.J. Bora, U.K. Saha, Experimental evaluation of a rice bran biodiesel-biogas run dual fuel diesel engine at varying compression ratios. Renew. Energy **87**, 782–790 (2016). https://doi.org/10.1016/j.renene.2015.11.002

64. B.J. Bora, U.K. Saha, Optimisation of injection timing and compression ratio of a raw biogas powered dual fuel diesel engine. Appl. Therm. Eng. **92**, 111–121 (2016). https://doi.org/10.1016/j.applthermaleng.2015.08.111

65. B.J. Bora, U.K. Saha, Theoretical performance limits of a biogas–diesel powered dual fuel diesel engine for different combinations of compression ratio and injection timing. ASCE J. Energy Eng. **142** (2016). https://doi.org/10.1061/(ASCE)EY.1943-7897.0000293

66. M. Talibi, P. Hellier, N. Ladommatos, Combustion and exhaust emission characteristics, and in-cylinder gas composition, of hydrogen enriched biogas mixtures in a diesel engine. Energy **124**, 397–412 (2017). https://doi.org/10.1016/j.energy.2017.02.070

67. S. Verma, L.M. Das, S.C. Kaushik, S.K. Tyagi, An experimental investigation of exergetic performance and emission characteristics of hydrogen supplemented biogas-diesel dual fuel engine. Int. J. Hydrogen Energy **43**, 2452–2468 (2018). https://doi.org/10.1016/j.ijhydene.2017.12.032

68. N. Khatri, K.K. Khatri, Hydrogen enrichment on diesel engine with biogas in dual fuel mode. Int. J. Hydrogen Energy **45**, 7128 7140 (2020). https://doi.org/10.1016/j.ijhydene.2019.12.167

69. K. Cacua, A. Amell, F. Cadavid, Effects of oxygen enriched air on the operation and performance of a diesel-biogas dual fuel engine. Biomass Bioenergy **45**, 159–167 (2012). https://doi.org/10.1016/j.biombioe.2012.06.003

70. M. Feroskhan, S. Ismail, M.G. Reddy, A. Sai Teja, Effects of charge preheating on the performance of a biogas-diesel dual fuel CI engine. Eng. Sci. Technol. Int. J. **21**, 330–337 (2018). https://doi.org/10.1016/j.jestch.2018.04.001

71. A. Sarkar, U.K. Saha, Role of global fuel-air equivalence ratio and preheating on the behaviour of a biogas driven dual fuel diesel engine. Fuel **232**, 743–754 (2018). https://doi.org/10.1016/j.fuel.2018.06.016

72. A.V. Prabhu, A. Avinash, K. Brindhadevi, A. Pugazhendhi, Performance and emission evaluation of dual fuel CI engine using preheated biogas-air mixture. Sci. Total Environ. **754**, 142389 (2021). https://doi.org/10.1016/j.scitotenv.2020.142389

73. M. Maizonnasse, J.S. Plante, D. Oh, C.B. Laflamme, Investigation of the degradation of a low-cost untreated biogas engine using preheated biogas with phase separation for electric power generation. Renew. Energy **55**, 501–513 (2013). https://doi.org/10.1016/j.renene.2013.01.006

74. I.D. Bedoya, A.A. Arrieta, F.J. Cadavid, Effects of mixing system and pilot fuel quality on diesel–biogas dual fuel engine performance. Biosour. Technol. **100**, 6624–6629 (2009). https://doi.org/10.1016/j.biortech.2009.07.052

75. N. Khayum, S. Anbarasu, S. Murugan, Combined effect of fuel injecting timing and nozzle opening pressure of a biogas-biodiesel fuelled diesel engine. Fuel **262**, 116505 (2020). https://doi.org/10.1016/j.fuel.2019.116505

76. I.T. Yilmaz, M. Gumus, Investigation of the effect of biogas on combustion and emissions of TBC diesel engine. Fuel **188**, 69–78 (2017). https://doi.org/10.1016/j.fuel.2016.10.034

77. F.Z. Aklouchea, K. Loubara, A. Bentebbiche, S. Awad, M. Tazerout, Predictive model of the diesel engine operating in dual-fuel mode fuelled with different gaseous fuels. Fuel **220**, 599–606 (2018). https://doi.org/10.1016/j.fuel.2018.02.053

78. R.G. Papagiannakis, D.T. Hountalas, C.D. Rakopoulos, Theoretical study of the effects of pilot fuel quantity and its injection timing on the performance and emissions of a dual fuel diesel engine. Energy Convers. Manage. **48**, 2951–2961 (2007). https://doi.org/10.1016/j.enconman.2007.07.003

79. X. Liu, H. Wang, Z. Zheng, M. Yao, Numerical investigation on the combustion and emission characteristics of a heavy-duty natural gas-diesel dual-fuel engine. Fuel **300**, 120098 (2021). https://doi.org/10.1016/j.fuel.2021.120998

80. U. Azimov, E. Tomita, N. Kawahara, Y. Harada, Premixed mixture ignition in the end-gas region (PREMIER) combustion in a natural gas dual-fuel engine: operating range and exhaust emissions. Int. J. Engine Res. **12**, 484–497 (2011). https://doi.org/10.1177/1468087411409664

81. U. Azimov, E. Tomita, N. Kawahara, Y. Harada, Effect of syngas composition on combustion and exhaust emission characteristics in a pilot-ignited dual-fuel engine operated in PREMIER combustion mode. Int. J. Hydrogen Energy **36**, 11985–11996 (2011). https://doi.org/10.1016/j.ijhydene.2011.04.192
82. U. Azimov, E. Tomita, N. Kawahara, Ignition, combustion and exhaust emission characteristics of micro-pilot ignited dual-fuel engine operated under PREMIER combustion mode. SAE Tech. Paper 2011-01-1764 (2011). https://doi.org/10.4271/2011-01-1764
83. C. Aksu, N. Kawahara, K. Tsuboi, S. Nanba, E. Tomita, M. Kondo, Effect of hydrogen concentration on engine performance, exhaust emissions and operation range of PREMIER combustion in a dual fuel gas engine using methane-hydrogen mixtures. SAE Tech. Paper 2015-01-1792 (2015). https://doi.org/10.4271/2015-01-1792
84. C. Aksu, N. Kawahara, K. Tsuboi, M. Kondo, E. Tomita, Extension of PREMIER combustion operation range using split micro pilot fuel injection in a dual fuel natural gas compression ignition engine: a performance-based and visual investigation. Fuel **185**, 243–253 (2016). https://doi.org/10.1016/j.fuel.2016.07.120
85. N. Kawahara, Y. Kim, H. Wadahama, K. Tsuboi, E. Tomita, Differences between PREMIER combustion in a natural gas spark-ignition engine and knocking with pressure oscillations. Proc. Combust. Inst. **37**, 4983–4991 (2019). https://doi.org/10.1016/j.proci.2018.08.055
86. A. Valipour Berenjestanaki, N. Kawahara, K. Tsuboi, E. Tomita, End-gas autoignition characteristics of PREMIER combustion in a pilot fuel-ignited dual-fuel biogas engine. Fuel **254**, 115634 (2019). https://doi.org/10.1016/j.fuel.2019.115634
87. A. Valipour Berenjestanaki, N. Kawahara, K. Tsuboi, E. Tomita, Performance, emissions and end-gas autoignition characteristics of PREMIER combustion in a pilot fuel-ignited dual-fuel biogas engine with various CO_2 ratios. Fuel **286**, 119330 (2021). https://doi.org/10.1016/j.fuel.2020.119330
88. E. Tomita, N. Kawahara, J. Zheng, Visualization of auto-ignition of end gas region without knock in a spark-ignition natural gas engine. J. KONES Powertrain Transp. **17**, 521–527 (2010). http://yadda.icm.edu.pl/yadda/element/bwmeta1.element.baztech-article-BUJ7-0018-0008

Chapter 4
Advanced Combustion Technologies for Higher Thermal Efficiency

Abstract Recently, new technologies for higher thermal efficiency have been developed in internal combustion engines. New types of combustion processes and technologies have been proposed regardless of the type of an engine. Low temperature combustion (LTC) such as homogeneous charge compression ignition (HCCI) has been studied to achieve low NOx (oxides of nitrogen) and particulate matter (PM). This chapter discusses history and advantages of the HCCI combustion. HCCI combustion technology has been evolved into the reactivity controlled compression ignition (RCCI) and spark assisted compression ignition (SACI) types. Abnormal knocking combustion that occurs due to the flame propagation and thermochemical autoignition in the end-gas region is considered one of the barriers to achieve higher thermal efficiency. However, under controlled conditions, knocking combustion can be avoided in SACI and in the Premixed Mixture Ignition in the End-gas Region (PREMIER) combustion processes. Furthermore, new laser- and plasma-based ignition systems have been developed instead of conventional spark ignition system. Laser ignition, non-thermal plasma assisted ignition such as microwave assisted spark ignition, nanosecond pulsed discharge, corona, etc. are described. These technologies are explained in this chapter, which shows future directions of internal combustion engines toward increasing thermal efficiency and minimizing NOx and PM emissions.

Keywords Internal combustion engine · Gas engine · Biogas · HCCI · RCCI · Low temperature combustion · Spark assisted compression ignition · Ignition · Laser ignition · Microwave assisted ignition · PREMIER combustion

Abbreviations

ATDC	After TDC
BGES	Biogas energy share
BMEP	Brake mean effective pressure
BTE	Brake thermal efficiency

© The Author(s), under exclusive license to Springer Nature Switzerland AG 2022   73
E. Tomita et al., *Biogas Combustion Engines for Green Energy Generation*,
SpringerBriefs in Applied Sciences and Technology,
https://doi.org/10.1007/978-3-030-94538-1_4

BTDC   Before TDC
CAD   Crank angle degree
CAI   Controlled auto ignition
CA50   50% Cumulative heat release
CDC   Conventional diesel combustion
CO   Carbon monoxide
DEE   Diethyl ether
DF-PCCI   Dual fuel PCCI
EGR   Exhaust gas recirculation
HC   Hydrocarbon
HCCI   Homogeneous charge compression ignition
IMEP   Indicated mean effective pressure
KI   Knock intensity
LTC   Low temperature combustion
MFB   Mass fraction burned
NOx   Oxides of nitrogen
NVO   Negative valve overlap
PCCI   Premixed charge compression ignition
PM   Particulate matter
PREMIER   Premixed mixture ignition in the end gas region
RCCI   Reactivity controlled compression ignition
RI   Ringing intensity
SACI   Spark assisted compression ignition
SPCCI   Spark controlled compression ignition
TDC   Top dead center
THC   Total hydrocarbons

4.1 Introduction

This chapter focuses on several advanced combustion techniques for higher thermal efficiency with concurrent control of exhaust gas emissions. Although these methods have been considered and developed mainly for gasoline and diesel engines, these concepts can be easily applied to biogas engines in the future.

A flame kernel is produced at a spark plug in an SI engine and flame propagates in a homogeneous mixture towards a cylinder wall. The subject that should be improved is the stable ignition, faster turbulent flame propagation, and end-gas combustion without knocking. Improvement of ignition source has been considered because the spark energy of an ordinary spark plug for gasoline engine is not suitable in lean and/or diluted mixtures and/or in strong gas flows because of instability of ignition and misfire. Higher spark energy using ordinary spark plugs is one of the solutions, however, durability of the tip of the spark plug may be a problem. Therefore, several new ignition systems have been considered instead of a conventional spark ignition, namely, laser ignition, and non-thermal plasma-assisted ignition.

On the other hand, utilizations of autoignition have been considered as initiation of the combustion in spark ignition engines. When fuel–air homogeneous mixture is compressed and required levels of temperature and pressure are reached, the mixture is autoignited. Then, the combustion proceeds entirely in the cylinder gradually and knocking does not occur under lean or diluted condition. Thermal efficiency is almost the same as that of diesel engine with ultra-low NO_x and PM. This concept is called homogeneous charge compression ignition (HCCI), which combines the features of gasoline and diesel engines. This combustion mode has been also tried in natural gas engines. However, stand-alone HCCI combustion has not yet been realized in commercial gas engines due to the challenges associated with the control of ignition.

In diesel engines, the initiation of the combustion is autoignition of mixture of injected fuel and air. However, the progressing spray combustion causes significant NOx and PM increases. Therefore, many concepts based on promotion of mixture homogeneity and low temperature combustion have been proposed to reduce NOx and PM simultaneously.

By adopting HCCI concept, spark assisted compression ignition (SACI) have been developed for spark ignition engines. In HCCI engine, the autoignition timing cannot be controlled when the engine speed and/or load changes. The autoignition timing is determined based on mixture temperature, pressure, and equivalence ratio. When the operating conditions are changing, the autoignition timing is not changing with crank angle but with time. Then the engine does not operate stably. Therefore, the concept of spark assisted HCCI have been proposed to control the ignition timing. The concept of reactivity controlled compression ignition (RCCI) has also been proposed in dual-fuel engines. High octane or methane number fuel is mixed well with air during compression and high cetane number fuel is injected before the compression TDC (top dead center). When the injection timing is much advanced compared to that in a normal diesel engine, the injected fuel is also mixed locally with surrounding fuel–air mixture. Then, the fuel is autoignited with many hot spot locations with mild combustion. For instance, the PREMIER combustion presented in Chap. 3 has been developed to utilize mild autoignition in the end-gas region in gas engines based on the concept of HCCI.

4.2 HCCI—Homogeneous Charge Compression Ignition

4.2.1 Concept of HCCI Combustion

While the term of HCCI combustion was proposed [1], CAI (controlled auto ignition) is often used as the same concept as HCCI [2]. Before the term of HCCI was proposed, some similar concepts have been proposed as ATAC (active thermo-atmospheric combustion) [3], TS (Toyota-Soken) [4], CIHC (compression ignited homogeneous charge) [5], the basic concept [6], etc. These concepts are based on compression ignition of homogeneous lean and/or diluted mixture in gasoline engines. Homogeneous lean fuel–air mixture in the cylinder is compressed and air–fuel mixture in the combustion chamber burns volumetrically. The advantages are (1) the achievement of higher thermal efficiency due to not using throttle valve that is used to reduce air flow to maintain stoichiometric fuel–air ratio at low loads, shorter duration of combustion and reduced heat transfer losses, and (2) the reduction of both exhaust emissions of PM and NO_x significantly due to lean homogeneous mixture and low combustion temperature, respectively. However, main challenges for HCCI combustion concept are the difficulty in combustion phasing control, strong noise, high level of unburned hydrocarbons (HC) and carbon monoxide (CO) emissions, limited operating range, and problems of cold start and homogeneous mixture preparation.

Thereafter, many research attempts have been made to resolve these disadvantages. Many review papers and books were also published on HCCI combustion and related combustion processes, and several examples were described in the following subsections. The overview of gasoline fueled HCCI, diesel fueled HCCI/CAI, the effect of fuel and modeling were reviewed [2, 7]. The overview of HCCI, advancements in the studies on HCCI fundamentals, evolution in control strategies of diesel and gasoline HCCI engines were summarized [8]. Performance and exhaust emission parameters of HCCI combustion were reviewed mainly for gasoline and gas engines [9]. The impacts of these tactics on ignition, combustion, and exhaust emissions for HCCI, LTC and other combustions were summarized, as well as the fuel design and management for advanced compression ignition combustion mode [10]. The fundamental factors impacting LTC and HCCI engines, as well as high load limitations and ways for exceeding them, were discussed [11]. Low temperature combustion was reviewed from the viewpoints of advantages and challenges for future engines [12, 13]. Some effective techniques and controlling strategies of the ignition timing and combustion phase in HCCI engine were reviewed [14]. The specifications of a test engine, its performance and emission behavior under the LTC mode were listed and discussed [15]. The principles of HCCI combustion, challenges of HCCI, and future directions were briefly summarized [16].

Different strategies and various challenges associated with the HCCI combustion were discussed and summarized, although clear solutions have not yet been obtained. There are still many challenges related to the development of robust ignition and combustion control strategies for HCCI technology. These include research on required exhaust gas recirculation (EGR) rate at different HCCI operating regimes, variation of fuel–air ratio at different loads, optimum compression ratio, inlet temperature and pressure of a fresh charge, fuel mixing strategies, optimum fuel characteristics, spark-assisted ignition strategies, engine cooling strategies, etc. These all were examined and debated. In spark ignition engines, load is controlled by throttling the intake air in order to maintain the same air–fuel ratio around the stoichiometric condition. In gasoline engines, the exhaust emissions of NO_x, HC, and CO are processed with three-way catalyst (TWC) to meet the regulation level under stoichiometric operation. When an engine operates under lean burn condition, such as HCCI, it is difficult for the TWC to reduce these emissions effectively. Therefore, one of the methods for solving this problem is to form diluted mixture in the cylinder by using hot and/or cold EGR under stoichiometric condition. Another method is to use oxidation catalyst under lean burn condition. To avoid strong noise due to rapid combustion, higher load should be restricted. Weak stratification of fuel–air and/or gas temperature leads to mild chemical reactions. Increasing gas temperature by hot EGR or inlet heating helps to operate at cold start.

At high loads, autoignition can be a source of high levels of combustion noise. The ringing intensity (RI, MW/m^2) is often used to evaluate the combustion noise of engine, which is defined as (4.1) [17]. This parameter is used even for non-knocking conditions.

$$RI = \frac{1}{2\kappa} \frac{\left(\beta\left(\frac{dP}{dt}\right)_{max}\right)^2}{P_{max}} \sqrt{\kappa R T_{max}} \qquad (4.1)$$

where, $(dP/dt)_{max}$ is the maximum pressure rise rate (kPa/ms), P_{max} is the peak in-cylinder average pressure, and T_{max} is the cylinder temperature.

The parameter β relates the amplitude of pressure fluctuations to the maximum pressure rise rate, which is set to 0.05. R is the gas constant and κ is the ratio of specific heats. Ringing intensity is limited to 2 MW/m^2 based on the acceptable combustion noise. However, later works used values of near 5 MW/m^2 [18–20]. Several research results show that boosted HCCI combustion can achieve 99% combustion efficiency, 47% thermal efficiency, with low NO$_x$ production [18], 54% indicated thermal efficiency, 16–18 bar IMEP (indicated mean effective pressure) with low NO$_x$ and soot under ~ 50% EGR [21]. At lower loads, cycle-to-cycle fluctuation was increased, and engine finally misfired. In all sources mentioned above, it was determined that combustion stability was limited by RI on the advanced range, and by cycle-to-cycle fluctuation of IMEP on the retarded range.

As described above, the HCCI engine has not yet been commercialized because these problems have not been resolved fully. The following sections discuss the potential for RCCI and SACI concepts to be applied in biogas engines.

4.2.2 Low Temperature Combustion in Diesel Engines

In diesel engines, new combustion technologies have been desired to reduce NOx and PM simultaneously other than aftertreatment of exhaust emissions. The concept of HCCI is homogeneous mixture and low temperature combustion. However, it is difficult to make a homogeneous mixture in a diesel engine. Then, by promoting stratification of fuel and air, the degree of homogeneity can be increased locally. As shown in Fig. 4.1, schematic diagram of new types of combustion modes are compared to the conventional diesel engine in relation to equivalence ratio ϕ and burned gas temperature T presented by researchers in Southwest Research Institute [22, 23]. At first, this $\phi - T$ map was proposed by Kamimoto and Bae [24] to explain the region of NO$_x$ and soot productions on $\phi - T$ map and was developed to explain the temporal changes of ϕ and T by Akihama et al. [25]. Soot and NO$_x$ regions

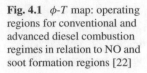

Fig. 4.1 ϕ-T map: operating regions for conventional and advanced diesel combustion regimes in relation to NO and soot formation regions [22]

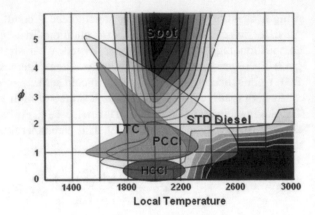

are presented in contour lines according to their concentrations. The region of HCCI combustion is on the left side near the bottom. HCCI combustion implies to maintain the equivalence ratio below 1.0 because the fuel–air mixture is very well premixed within the long induction time, and the combustion temperature is relatively lower than that in NO_x formation region.

After the HCCI was proposed, some new combustion methods were also proposed. UNIBUS (uniform bulky combustion systems) has a centrally mounted pintle-type injector, and early injection around 60°BTDC (before TDC) promotes mixing of fuel and air, achieving both low NO_x and low smoke [26]. The strategy of MK (modulated kinetics) combustion had retarded fuel injection with high EGR rate, and the same timing of autoignition and the end of injection to realize lean and partially premixed mixture formation [27]. On the other hand, PCCI (premixed charge compression ignition) combustion was firstly proposed for gasoline engine [28] and thereafter mainly applied to improve the diesel combustion. In this concept, some stratification is permitted with premixed fuel–air mixture in a real engine. These are summarized in a chapter of textbook [29]. The concept is based on low temperature combustion, avoiding soot and NO_x regions in $\phi - T$ map shown in Fig. 4.1. Various names other than the above were proposed, mainly for diesel engines, that is, PREDIC (premixed lean diesel combustion) [30], PCI (premixed compression ignition) [31], LTRC (low temperature fuel rich combustion) [29], LTC [32], etc. Thereafter, high efficiency clean combustion (HECC) for LTC, which showed high combustion efficiency (~100%), and near zero NOx and soot emissions was proposed [33].

The use of early, multiple, and late injection techniques to accomplish HCCI combustion in diesel engines was examined [34]. Under partially premixed conditions, the LTC of a diesel engine was investigated [35]. Early injection strategies were reviewed and discussed for HCCI and PCCI combustion in diesel engines [36].

4.2.3 HCCI Combustion in Biogas Engines

Biogas, which is composed of roughly 60% CH_4 and 40% CO_2, is even more difficult to be autoignited because ignition delay is longer due to slow chemical reaction with CO_2 as shown in Fig. 2.14 and unburned gas temperature becomes lower during compression stroke due to high heating capacity. When biogas of 67% CH_4 and 33% CO_2 was used as fuel in an HCCI engine with compression ratio of 15, very high inlet gas temperature of 250 °C was required to sustain combustion compared to other gaseous fuels [37].

HCCI combustion with biogas and diesel fuels was investigated [38, 39]. Diesel fuel was injected into an intake manifold to mix with biogas and air. The intake temperature was changed from 80 to 135 °C to ignite the mixture near TDC. The engine was operated at BMEP (brake mean effective pressure) = 2.5–4.0 bar with changing the ratio of biogas to diesel. The results showed lower thermal efficiency, NO_x and smoke, and higher HC compared to straight diesel. The engine operations of CI mode below BMEP = 2.5 bar, dual-fuel mode above BMEP = 4.0 bar, and HCCI mode between the two BMEPs were recommended.

An engine was modified for HCCI combustion and biogas was fueled in cylinder 2 and 3, while gasoline was fueled in cylinder 1 and 4 to ensure that the engine block temperature is maintained at required level [19, 20, 40]. Researchers investigated under the conditions of inlet pressure P_{in} = 2.0–2.2 bar (abs.), inlet temperature T_{in} = 473–483 K (200–210 °C), ϕ = 0.25–0.4 with biogas of 60% CH_4 and 40% CO_2. Figure 4.2 displays in-cylinder pressure, gross rate of heat release, and cyclic variability in the peak of gross heat release rates for various inlet absolute pressures, equivalence ratios, and CA50 with increased loads. Figure 4.2a shows that the delayed CA50 made peak pressure lower due to the decrease in the rate of pressure rise. When combustion was delayed, rates of pressure rise became smaller because overall burning rates and gas temperatures decreased. As shown in Fig. 4.2b, when CA50 was delayed, the peak in gross heat release rate was also delayed and decreased. At the timing of CA50 corresponded to 4 or 5 crank angle degrees (CAD), the ringing intensity has reached almost the limit. Figure 4.2c presents that the peak of the gross heat release rates showed cycle-to-cycle variations slightly when CA50 was advanced until 10°ATDC. However, when CA50 reached 12°ATDC, cycle-to-cycle fluctuation became larger, causing lower indicated thermal efficiency and $IMEP_g$ as well as higher HC and CO emissions, which are not presented here. Even when inlet conditions changed slightly, the cyclic fluctuations showed largely at lower equivalence ratios and the ringing intensity became large at higher equivalence ratios. At lower loads, gasoline pilot injection into an intake port has improved the cycle-to-cycle fluctuation. Researchers also simulated the biogas combustion with 12-zone model to evaluate the effects of CH_4 concentration (30–80%) and oxygen ratio (21, 22, and 27%) [41]. It was discovered that the key benefits of employing biogas in HCCI mode in CHP (combined heat and power) systems was an improved efficiency and reduced emissions.

Fig. 4.2 (**a**) In-cylinder
pressure. (**b**) Gross heat
release rate at inlet pressures
(abs.) of 2 and 2.2 bar, $\phi =$
0.4 and 0.5, for 50%
cumulative heat release
(CA50). (**c**) Cycle-to-cycle
fluctuations in the peak of
ROHR with plotted average
values as dashed bold lines
[20]

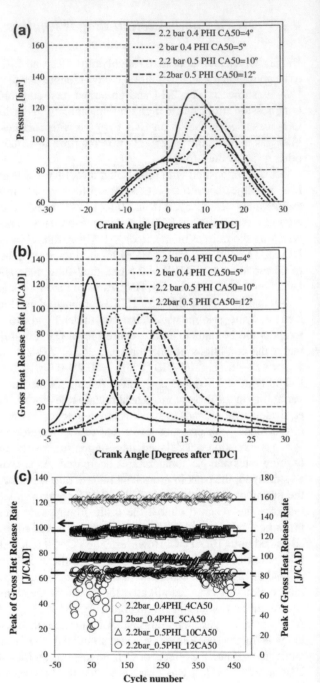

Fig. 4.2 (**a**) In-cylinder pressure. (**b**) Gross heat release rate at inlet pressures (abs.) of 2 and 2.2 bar, $\phi =$ 0.4 and 0.5, for 50% cumulative heat release (CA50). (**c**) Cycle-to-cycle fluctuations in the peak of ROHR with plotted average values as dashed bold lines [20]

The effect of negative valve overlap (NVO) on the change of the gas temperature in the cylinder in biogas fueled HCCI was investigated numerically [42]. Internal EGR caused the temperature in the cylinder to rise. When n-heptane was used as a high reactivity ignition promoter that was injected into the intake port with biogas and air, moderate amount of n-heptane did not change the combustion efficiency and HC emissions, and led to higher IMEP, and gross indicated efficiency [43].

Diethyl ether (DEE) was mixed in biogas as an ignition improver because the cetane number of DEE is about 125 [44]. It was found that brake thermal efficiency (BTE) in HCCI was larger than those in biogas-diesel dual-fuel mode and biogas SI mode. In HCCI mode, NO_x and smoke were decreased and HC was lower than that in lower loads and the same at 100% load in SI mode. The increasing DEE mixing with biogas and lower biogas flow rate offered improved BTE and lower CO and HC [45]. Higher degree of homogeneity of DEE with biogas that was injected into an intake manifold showed higher BTE compared to intake port injection [46].

A simulation with single zone model for biogas HCCI mode was performed with a small amount of diethyl ether injected into a manifold during the intake stroke [47]. A one-zone and GRI-mech chemical reaction mechanism was also applied for CH_4 and H_2 blends in HCCI mode [48]. Another simulation for biogas HCCI mode was performed with newly developed CANTERA reaction flow package [49]. The findings showed that increasing the equivalence ratio over 0.25 considerably increases NOx emissions by increasing the bulk temperature at combustion peak. However, when the equivalence ratio went above 0.25, CO and unburned hydrocarbons (HC) emissions reduced dramatically, but they only altered marginally when the CH_4 ratio and inlet temperature increased.

Possibility of biogas with hydrogen combustion was investigated in a controlled autoignition engine. Simulations were performed for CH_4 and CO_2 with H_2 and hot EGR, keeping the same IMEP of 8.4 bar at $\phi = 0.4$ [50, 51]. The hydrogen addition to biogas reduced autoignition temperature, so the biogas required less hot EGR that was used to increase the charge temperature. Although the boost pressure was reduced, the thermal efficiency did not change significantly. Therefore, lower boost pressure and smaller EGR resulted in lower NO_x emissions at moderate hydrogen ratio (CH_4: CO_2: H_2 = 0.581: 0.355: 0.065).

4.3 RCCI—Reactivity Controlled Compression Ignition

When a high cetane fuel such as diesel fuel was injected at early timings around 30–60°BTDC into the cylinder with gaseous fuel (methane, natural gas or iso-octane) with high methane number and air mixture in dual-fuel engines, the combustion became mild and showed higher thermal efficiency with low NO_x and low smoke (e.g., [52–55]). High cetane number fuel is roughly mixed with high methane number fuel and fuel stratifications of ignitability is achieved, so that it is considered that ignition occurs at many locations after long ignition delay times. The disadvantage was higher HC and CO emissions. The name of RCCI (reactivity controlled compression ignition) was proposed for this kind of combustion [56], which was based on HCCI combustion with early injection of diesel fuel. Researchers showed both results of experiment and simulation for NOx, soot, gross indicated thermal efficiency and ringing intensity over the range of 4.6–14.6 bar $IMEP_g$ at engine speed of 1300 rpm. Gasoline and diesel fuels in the experiments, and i-octane and n-heptane in simulation, were used as dual fuel approaches, NOx and soot levels were very low. The most superior thermal efficiency was 56% at 9 bar IMEP with acceptable ringing intensity. The simulation results agreed very well with those in experiments. The timings of the injection start in the cylinder were 58° and 37°BTDC for squish area and for piston bowl, respectively. The ROHR of RCCI presented a small peak near 10°BTDC due to the low-temperature oxidation, and a higher peak near 5°ATDC due to the main combustion.

Some review papers reporting on RCCI research were published [57–59]. In RCCI, two fuels with different reactivities are used to control the combustion phasing. High methane number or octane number fuel is supplied through the intake port or manifold during the intake stroke. High cetane number fuel is injected at high pressure directly into the cylinder at early timing compared to normal diesel operation. Also, this strategy controls the difference in combustion time between two fuels due to spatial stratification. In comparison to traditional diesel combustion, the dual-fuel RCCI produces ultra-low NOx and soot emissions and has a better thermal efficiency. It was also shown that the RCCI engine has a potential for achieving 60% indicated thermal efficiency [60].

Figure 4.3 shows the engine performance and exhaust gas emissions for conventional diesel combustion (CDC), PCCI, HCCI and dual-fuel PCCI (DF-PCCI), using the same engine fueled with CNG and diesel at low load (IMEP = 0.45 MPa) [61]. The DF-PCCI combustion is the same as RCCI combustion. The indicated thermal efficiency was the highest among those mentioned modes. The NOx levels were extremely reduced in PCCI, HCCI and DF-PCCI mode compared to the diesel only combustion. The PM in DF-PCCI mode showed lower than those in CDC, PCCI, and HCCI modes. The total hydrocarbons (THC) and CO in DF-PCCI mode were smaller than those in PCCI and HCCI, although THC and CO were larger than those in CDC mode.

The RCCI concept was also applied in biogas engine combustion. The early injection timings of 55–70°BTDC of diesel fuel was performed with a biogas induction engine [62]. Although the intake charge temperature needed 50–90 °C and the BMEP was limited between 2 and 4 bar, the results showed better BTE, significantly lower HC and NO_x compared to those of dual-fuel mode. It was found that diesel mode and dual-fuel mode were suitable in BMEP below 2 bars and BMEP above 4 bars, respectively. The RCCI combustion with biogas was investigated with injection timing of diesel fuel at 40°BTDC [63]. NO_x and soot were decreased while HC and CO were increased. The increase in ratio of biogas reduced NO_x level. The effect of hydrogen addition to landfill gas (50% CH_4 and 50% CO_2) was investigated using CFD (computational fluid dynamics) simulation with early injection of n-heptane at 35–51°BTDC [64]. EGR was also applied according to the added amount of hydrogen. Higher H_2 addition and EGR reduced NO_x. The effect of CH_4 concentration was investigated in

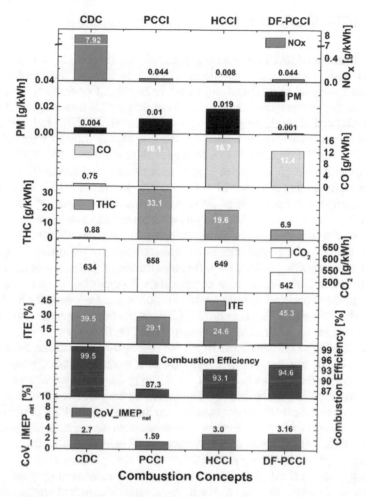

Fig. 4.3 Comparison of engine performance and exhaust emissions for CDC, PCCI, HCCI and DF-PCCI modes [61]

predominantly premixed charge compression ignition combustion, that was almost the same as RCCI combustion [65]. Injection timing was controlled at 33–91°BTDC according to CH_4 concentration and biogas energy share (BGES) ratio. Low proportions of CH_4 around 22–25% in biogas were more suitable for operation, leading to higher BMEP and lower smoke level. An engine was operated at 6.5 bar IMEP and 1600 rpm in biogas-diesel RCCI mode with changing CO_2 content (25–45%) and port mixing distance (0–115 mm from the intake valve) [66]. It was shown that the highest indicated thermal efficiency was obtained at 45% CO_2 content and 57.5 mm apart from the intake valve due to lower heat loss. Another RCCI engine was operated with abundant biogas-diesel, and NO_x was reduced by 75% with varying CO_2 content from 0 to 40% [67].

4.4 SACI—Spark Assisted Compression Ignition

The development of SACI combustion was reviewed [68]. If HCCI mode is used in a spark ignition engine, it would not be possible to cover all the operating range for engine speed and load. Therefore, SI mode and HCCI mode must be switched near the border of operating ranges. The transition region between HCCI and SI were investigated using the method of spark-assisted HCCI [69]. Both the air–fuel ratio and the compression ratio must be reduced significantly to transition from spark-assisted HCCI to purely SI combustion. The air–fuel ratio for HCCI combustion is virtually too lean to support flame propagation at the intermediate working stages, but the air–fuel ratio for SI combustion is too rich for HCCI combustion with the increased noise and large NOx emissions. At the boundary of autoignition, considerable cycle-to-cycle variations were predicted, since some cycles would attain autoignition conditions in the cylinder while others would not. As a result, wherever feasible, the engine should be operated at these intermediate operating points to avoid misfiring or knocking. Because the rise in pressure and temperature due to the initial flame propagation compensates for the pressure and temperature increase from compression, the compression ratio for spark-assisted HCCI cases will be somewhat lower than that for pure HCCI cases. In homogeneous mixtures, it is difficult to realize a smooth change between SI and HCCI modes.

The utilization of spark-assisted HCCI for stratified mixture in the cylinder was proposed [70]. It was reported the injection of a small amount of gasoline near the TDC, where flame propagation occurred due to the premixed mixture beyond flammability limit near the spark plug. Thereafter, the temperature of the mixture and pressure increased due to flame propagation and induced autoignition of surrounding homogeneous lean mixture. Therefore, pre-combustion temperature that is required to achieve HCCI was reduced, and the engine could be operated at higher loads up to 6.5 bar IMEP below the ringing limit.

Recently, Mazda © developed a spark controlled compression ignition (SPCCI) engine for production vehicles, of which, the combustion method was presented in 2019 [71]. There are two SPCCI operating regions: one is at A/F lean condition for light load and another is G/F lean condition that contains large amount of EGR for middle load. At very high loads and/or very high engine speeds, the combustion changes to SI mode. Figure 4.4 shows the rate of heat release under the condition of engine speed of 1500 rpm and IMEP = 500 kPa in a single cylinder test engine. Even if intake temperature, which is one of the causes of fluctuations, changed from base line to +15, +21 and +36 °C, the engine operated smoothly by adjusting the injection timing from base line (26°BTDC) to 23, 18 and 17°BTDC.

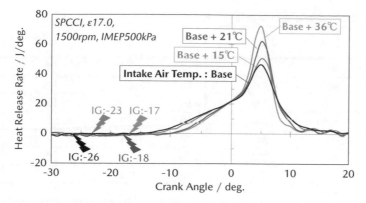

Fig. 4.4 ROHR of SPCCI combustion for different inlet temperature [71]

4.5 PREMIER—Premixed Mixture Ignition in the End-Gas Region

As presented in Chap. 3, PREMIER combustion is one of the methods for improving thermal efficiency for dual-fuel engines. It stipulates the advancement of the pilot injection timing of the liquid fuel to obtain one more peak in the rate of heat release during the latter part of combustion as presented in Figs. 3.9 and 3.10. Figure 4.5

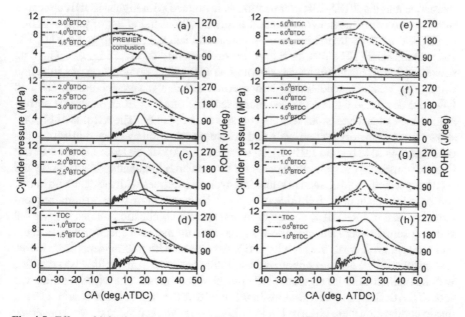

Fig. 4.5 Effect of injection timing on in-cylinder pressure and ROHR under various conditions indicated in Table 4.1 [72]

Table 4.1 Experimental conditions in Fig. 4.5

	P_{inj} MPa	P_{in} kPa	D_{hole} mm	N_{hole}	m_{DF} mg/cycle
(a)	40	200	0.10	3	3
(b)	80	200	0.10	3	3
(c)	120	200	0.10	3	3
(d)	150	200	0.10	3	3
(e)	40	200	0.10	3	2
(f)	80	200	0.10	3	2
(g)	150	200	0.08	3	3
(h)	150	200	0.10	4	3

P_{inj}: Pilot fuel injection pressure
P_{in}: Intake pressure
D_{hole}: Nozzle diameter
N_{hole}: Number of nozzles
m_{DF}: Pilot fuel injection mass per cycle

shows several examples of in-cylinder pressure and rate of heat release for various experimental conditions as shown in Table 4.1 [72]. The parameters were injection pressure P_{inj}, nozzle diameter D_{hole}, number of nozzle holes N_{hole} and injection mass per cycle m_{DF} at intake pressure $P_{in} = 200$ kPa and engine speed $= 1000$ rpm, fueled with natural gas. With the advancement of injection timing, the maximum pressure increases, and the PREMIER combustion is instigated with higher ROHR appeared in all shown experimental conditions. Figure 4.6 shows the effect of pilot injection timing and EGR on IMEP, indicated thermal efficiency η_i, as well as on NO_x, HC and CO emissions. The data in Fig. 4.6a correspond to those in Fig. 4.5. The data surrounded with black circles are denoted as PREMIER combustion. As injection timing θ_{inj} was advanced, IMEP and thermal efficiency were increased. HC and CO were decreased while NO_x was increased. The effect of EGR on engine performance and exhaust emissions was also presented in Fig. 4.6b. The left side of Fig. 4.7 shows pressure history for one cycle and pressure oscillation by processed with band-pass filter (4–20 kHz) into PREMIER, transient and knocking cycles. The FFT (fast Fourier transform) analysis of the pressure showed high frequencies of ~ 6.5 kHz and ~ 10 kHz, which could be seen in knocking cycles and agrees with the data with knock sensor as presented in the right side of Fig. 4.7. However, there were no high frequencies observed in PREMIER combustion and normal combustion modes. Figure 4.8 shows ensemble-averaged and time-resolved measurements of spectroscopy near the end-gas region. A small sensor whose tip was brazed with a small sapphire window was installed at the end-gas region in the cylinder head and a quartz fiber was coupled with intensified CCD (charge-coupled device) spectrometer. The exposure time was set to 3 CAD. Strong light emissions in ultraviolet wavelength of ~ 286 and ~ 310 nm due to combustion were observed in knocking cycles while no signals were observed in conventional (normal) cycles. These emissions are due to OH radicals that occur

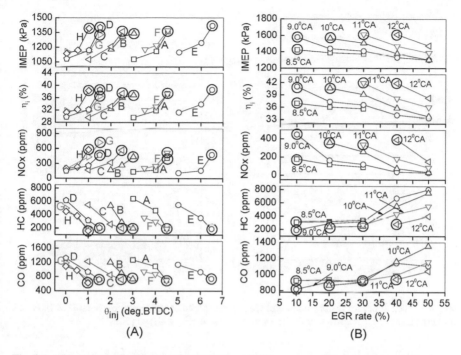

Fig. 4.6 Effect of injection timing of pilot fuel and EGR rate on engine performance and exhaust emissions [72]

during the auto-ignition of the mixture. On the other hand, these emissions could be also observed in PREMIER combustion, however, the intensities were weaker than those during knocking combustion. Therefore, autoignition in the end-gas regions occurs in PREMIER combustion without any pressure oscillations. The studies of PREMIER combustion were successively performed in dual-fuel engines fueled with methane-hydrogen mixture [73], natural gas [72, 74, 75], biogas [76, 77], and syngas [78]. The combustion visualization studies of the end-gas region autoignition in spark ignition engines were also performed [79, 80].

4.6 Advanced Ignition Systems

The mechanism of spark discharge of the transistorized coil ignition is briefly described in Sect. 2.2.1. In spark ignition engines, it is difficult to operate in leaner, more dilute, higher pressure, and stronger gas flow conditions by using a conventional spark plug system due to the expected increases in cycle-to-cycle fluctuations. To achieve higher thermal efficiency, the conditions in the cylinder must undergo through strict control and processes. Various ignition techniques have been developed and tested to alleviate combustion instability. High levels of energy supplied

88

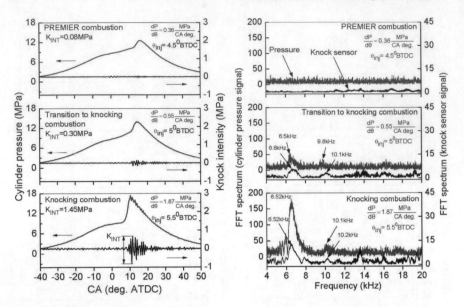

Fig. 4.7 Left: Pressure history and amplitude of the pressure processed by band-pass-filter (4–20 kHz); Right: FFT spectrum of the pressure and knock sensor signal [72]

to the spark plug are effective to produce a flame kernel, however, durability is not guaranteed due to the degradation and erosion of the tip of the spark plug. Therefore, enhanced systems for more reliable ignition have been investigated as alternatives to the conventional spark plug systems. Several high energy spark systems and enhanced ignition systems were summarized [81]. On the other hand, fundamentals, and wide range of applications of plasma assisted ignition and combustion were reviewed [82–86].

4.6.1   Laser Ignition

Laser ignition is achieved by focusing laser energy beyond the breakdown energy of the fuel–air mixture. The phenomena of the laser ignition have been studied by many researchers. At first, laser ignition was tried for various fuels in a constant volume vessel where the lower energy was needed under higher pressure conditions [87]. A CO_2 laser with strong power (300 mJ) was used to introduce the laser beam to an internal combustion engine with a single cylinder [88, 89]. The results showed that thermal efficiency and NO_x were increased, and the lean limit was extended up to A/F = 22.5. HC and CO were similar compared to those with conventional electrical spark plug. The spark location apart from the cylinder wall was effective. The properties of laser ignition were summarized theoretically with some experimental data [90]. Almost all engine results showed that stable combustion with laser spark plug

Fig. 4.8 Time-resolved measurement of spectroscopy of OH radicals near end-gas region [72]

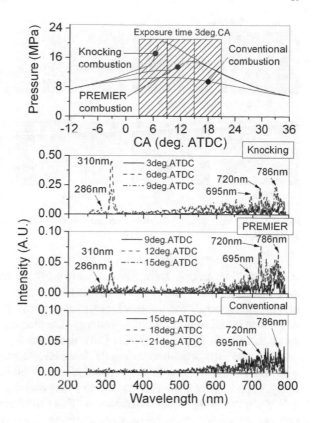

system was achieved under lean conditions and that the emissions of HC and CO were comparable to those with the electrical spark plug, while NO_x emissions were contradictory. The detail processes of formation of the plasma and ignition with laser were reviewed in terms of engine combustion and interpreted theoretically [91–94]. Detailed processes of formation of the plasma are described in these reviews.

Figure 4.9 shows visualization of laser ignition process of stoichiometric propane-air mixture with ultra-high-speed Schlieren method near the burner exit from laser shot till formation of stable burner flame [95]. A beam emitted from the left side with a second harmonic Nd:YAG laser was focused at the central position of the figure. At time $t = 1$ μs after the laser shot, a small plasma kernel was seen. At $t = 5$ μs, the hot kernel grew gradually, and a shock wave was produced. At $t = 11$ μs, the hot kernel became spherical. After $t = 120$ μs, the third lobe grew largely as shown in the results by Bradley et al. [91] who explained that the third lobe was created by gas dynamics around the kernel. The resulting rarefaction wave forms toroidal rings at the plasma's leading and trailing edges. The former decay more quickly, resulting in the formation of a third lobe of the kernel that travels toward the laser. When a flame is produced, the third lobe gradually fades away as the first gas dynamics effects fade. Finally, a burner flame was formed at $t = 49.8$ ms in this case.

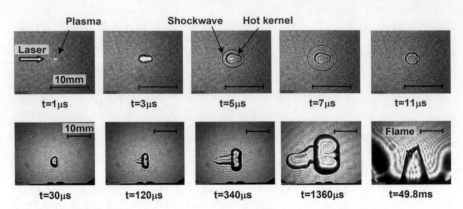

Fig. 4.9 Ultra-high speed Schlieren photographs of formation of the kernel produced by laser ignition [95]

The ignition process from the breakdown till the flame kernel formation was investigated with a spectroscopic analysis [96]. Figure 4.10 shows the evolution of the emission intensity for different incident energies of 50, 110 and 170 mJ, which are larger than the minimum breakdown energy. The progression of distinct wavelengths corresponding to H (656 nm), CN (387 nm), and N_2 (387 nm) was shown. The quantity of energy deposited during the laser breakdown instigated the temperature rise. Some atomic emissions were recorded due to full ionization in the plasma. For a larger spark energy, the molecular emissions remained longer, implying that the radicals in the flame kernel lasted longer. This time frame is crucial because a self-sustaining chemical reaction requires attaining a particular quantity of radicals at the proper moment and in a way that triggers branching phenomena. To investigate

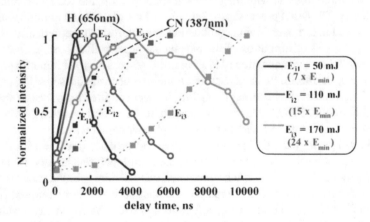

Fig. 4.10 H (656 nm), CN, N_2 (387 nm) emission intensities for different incident energies in the flame kernel (C_3H_8-air, $\phi = 1$; flow speed = 106 cm/s; $E_{in} = 50$ mJ = $7 \times E_{min}$; $E_{in} = 110$ mJ = $15 \times E_{min}$; $E_{in} = 170$ mJ = $24 \times E_{min}$) [96]

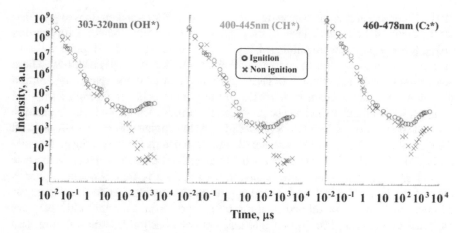

Fig. 4.11 Emission intensities for OH* (303–320 nm), CH* (400–445 nm) and C_2* (460–478 nm) for ignition and non-ignition (C_3H_8-air, $\phi = 1$; flow speed = 106 cm/s) [96]

the pivotal point, which is regarded as the beginning of the branching phenomenon, the intensity behaviors of the three wavelength ranges for OH^* (303–320 nm), CH^* (400–445 nm) and C_2^* (460–478 nm) with time are presented in Fig. 4.11. These three radicals are considered to be representative of premixed flame showing almost the same trend. A clear difference appeared between 20 and 300 μs. Each emission intensity decreased faster in the case of misfiring than in the case of firing. A slope discrepancy was observed around 20 μs. The curve for the case that fired was flatter than the curve for the case that misfired, which had a bigger negative slope.

It has been confirmed that a large laser with large energy such as Nd:YAG laser can be used as an energy source of ignition. In a direct injection engine, the gradient of fuel concentration near the spark plug increases due to stratification of the fuel. Then, adjustable ignition location is highly expected. A strong laser light must be introduced into an engine cylinder through an optical window and optical fiber. The window usually becomes dirty due to lubricant oil and soot, so that durability of the window must be guaranteed for many years to ensure effective system operation. A large-size laser cannot be used for automobile engines even if the principle of laser ignition is confirmed. In stationary gas engines, the size of the total system is not so critical as to compare to the size of automobile engines.

A four cylinder engine with a laser ignition system was fully operated with a Q-switched Nd:YAG laser on an optical bench [97]. Instead of a large laser on a table, a compact solid-state laser system was proposed [98]. A passively Q-switched solid-state laser, a Nd:YAG/Cr^{4+}:YAG laser end-pumped by a fiber-coupled laser diode was developed [99]. Based on this, laser spark plug prototypes were built by several groups. Instead of a single high-energy laser pulse, a pulse train of low-energy pulses was used to ignite a passively Q-switched Nd:YAG/Cr:YAG micro-laser [100], as was applied to a GDI (gasoline direct injection) automobile engine on a Toyota car

[101]. Other researchers also investigated various conditions and engines (e.g., [102–107]). Figure 4.12 shows a laser spark plug with three points locations for ignition [107]. Review papers were published, especially regarding the development of laser spark plug systems applied to internal combustion engines [94, 104, 108, 109]. These studies summarized advantages of laser ignition compared to the spark plug ignition as follows: (1) Laser beam is delivered in the engine cylinder through an optical window, so that flame kernel can be produced at various locations with no quenching because there is no spark electrode; consequently, the engine efficiency increases and the emissions of HC and CO decrease; (2) Multiple beams can be produced to ignite multiple locations at the same time or sequentially; (3) It is easier to ignite the mixture under higher pressure conditions, although the required voltage becomes higher in the electric spark plug systems. As the engine can be operated under the conditions of boosted, lean and/or diluted mixture, output power and thermal efficiency are expected to increase. Multi-point ignition, optical fiber guided transmission, and small laser and laser control systems have also been investigated to apply in road vehicles.

The laser ignition system was applied to natural gas engines [110]. Nd:YAG micro lasers that are passively Q-switched and have been designed for the steady operation under demanding engine conditions were mounted in the individual cylinders of a 350 kW lean-burn inline six-cylinder, open-chamber, spark-ignited engine, and testing were carried out at high-load (298 kW) and rated speed (1800 rpm) [111]. This stationary gas engine for power generation has comparatively large size displacement and high pressure in the cylinder. Because of the higher cost its use can be justified

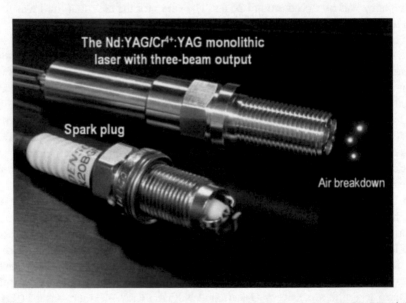

Fig. 4.12 A photo of the composite, all-ceramics passively Q-switched Nd:YAG/Cr4+:YAG monolithic laser with three-beam output (Reprinted with permission from [107] © The Optical Society)

compared to automobile engines. Laser ignition is more acceptable to be utilized in gas engines for power generation if the durability problems are resolved.

4.6.2 Non-thermal Plasma-Assisted Ignition System

Due to the ability to rapidly generate active radicals and exciting species by electron impact dissociation, excitation, and subsequent energy relaxation, non-thermal plasma became an interesting solution to provide higher electron temperatures and kinetic activity [85]. In this section, microwave, nano second discharge, and corona discharge are explained briefly.

It is well known that microwave affects the behavior of premixed flame [112–114]. The microwave can enhance ignition stability for gas turbines, scram jets, internal combustion engines, and the basic phenomena have been discussed [115]. The microwave-assisted spark plug was developed for a gasoline engine [116–118], which initiated plasma using a conventional spark discharge, with enhanced electron energy and expanded plasma by emitting microwaves into the spark zone. The microwave-assisted spark plug does not create plasma without first generating a spark discharge, indicating that plasma is not formed solely by the sub-critical microwaves causing a coronal discharge between the conducting spark plug electrode and the ground. Initial combustion duration became shorter because much OH radicals produced near the spark plug volumetrically enhanced the formation of the flame kernel under lean conditions. Then, cycle-to-cycle fluctuations of the mean effective pressure were reduced, and the engine operated stably. The lean limit was also extended from $\phi = 0.76$–0.61. Almost the same results were obtained for other engines [119, 120]. The performance of microwave-assisted spark plug technology was investigated for different methane–air combinations at a variety of starting pressures and microwave energy inputs in a constant volume combustion chamber [121]. The microwave-assisted spark ignition shortened initial combustion duration, 0–10% MFB (mass fraction burned), at all equivalence ratios and extended the lean limit to $\phi = 0.57$, however, did not affect main combustion duration, 10–90% MFB. The microwave discharge igniter that can generate plasma in air and CO_2 up to 6 MPa was developed [122]. A multiple microwave discharge igniters with three ports in a plug were developed and tested in a spark ignition engine [123]. When all three ports created the discharge simultaneously, the lean limit was extended and 0–10% MFB became shorter under lean conditions.

Nanosecond pulsed discharge has also been performed in a spark ignition engine. High voltage brief pulses (20–85 ns) with energy (60 mJ) equivalent to that of a standard spark igniter (2 ms, 80 mJ) were used to create the transient plasma [124]. The plasma igniter can produce reactive species on nanosecond time scales by electron impact dissociation of fuel and air mixture, resulting in a vast volume of scattered arrays of streamers. The ignition delay time was found to be improved when the plasma igniter was used. Stable lean combustion was achieved at conditions not ignitable by a spark plug when a 20 ns pulse was utilized. It was also discovered that utilizing the 20 ns pulse, it was possible to achieve the peak pressure 20% higher than that of spark ignition due to faster flame speeds compared to those of the spark ignition [125–127]. Ignition properties in the lean combustion area, as well as the link between pulse width and ignition characteristics, were examined further. With a fixed input energy of 60 mJ, the plasma igniter had a pulse width of 80 ns or 25 ns. Because the arc transition was made more difficult by reducing the pulse width from 80 to 25 ns, the maximum voltage of the streamer discharge was increased at the same pressure conditions, resulting in a 30–40% increase in the reduced electric field strength E/N, where E is the electric field strength and N is the gaseous molecular number density. The findings showed that a shorter pulse width resulted in quicker combustion. The low temperature plasma means that the plasma temperature is high, but the gas temperature is not so high under the non-equilibrium conditions. Therefore, the heat loss of plasma to the electrodes is reduced. The lower gas temperature leads to production of lower level of nitrogen oxides. Similar concept of repetitive 10–100 ns pulse discharge systems was proposed [128–130]. Similarly, high frequency ignition system was also proposed by Delphi © to be applied in a stratified combustion engine [131, 132]. Researchers installed a low-temperature plasma igniter (barrier discharge igniter (BDI)) with high-frequency voltage (15 kHz) in the upper center of the combustion chamber [133]. It was discovered that the autoignition characteristics of the HCCI combustion in gasoline engine by using this igniter were superior to those of spark-assisted HCCI. This allowed an expansion of the zones of stable HCCI operation on the lean mixture side and on the low intake air temperature side by using a single-cylinder gasoline engine.

A high frequency continuous plasma ignition system was tested in a multi-cylinder gasoline engine [134]. A 5-star shaped electrode ignitor and a resonant transformer circuit was used. With a frequency of 4.697 MHz, the input voltage of 100 V was amplified up to 6 kV. The corona discharge at the end of each sharp electrode caused the ignition. When compared to a typical spark plug system, the engine test results revealed that fuel economy and combustion stability were enhanced. Although CO and HC emissions were reduced, the greater in-cylinder temperature resulted in large increase in NOx emissions.

4.7 Others

For conventional internal combustion engines, an increase in compression ratio and lean burn can lead to higher thermal efficiency according to the equation of thermal efficiency of Otto cycle [135]. In spark ignition engines, reduction of pumping losses at low loads is important. Supercharged or turbocharged intake systems with Miller cycle have been often used to increase output power and thermal efficiency with suppressed knocking. Miller cycle was developed and patented for the improvement of output power in supercharged engines without knock by substantial reduction of compression ratio but maintaining expansion ratio owing to varying the timing of the intake valve closure [136]. For Miller cycle, the implementation of variable valve timing for opening and closing, or effective compression (expansion) ratio is desirable. A project of Innovative Combustion Technology was organized in the Cross-ministerial Strategic Innovation Promotion Program (SIP) by the Cabinet Office in Japan recently. They achieved 51.5% thermal efficiency in a test spark ignition engine with super lean combustion owing to high ignition energy, strong tumble motion, etc. and 50.1% thermal efficiency in a diesel engine in 2019 [137].

When the power generation engines operate at the best performance conditions, cycle-to-cycle and cylinder-to-cylinder variations must be suppressed as much as possible. This is because only one cycle of knocking or misfire can affect the following cycle. Therefore, it would be very important to precisely control the in-cylinder combustion, and to achieve that the robust engine control systems would be required.

4.8 Summary

This chapter has covered advanced combustion technologies, which can be applied to internal combustion engines fueled with biogas. Some techniques have not yet been investigated with biogas engines. However, these techniques can be easily applied to biogas, sygas and hydrogen gas engines. Biogas fueled engines are expected to achieve higher thermal efficiency and lower exhaust gas emissions for wider applications in the future.

References

1. R.H. Thring, Homogeneous-Charge Compression-Ignition (HCCI) Engines. SAE Tech. Paper 892068 (1989). https://doi.org/10.4271/892068
2. H. Zhao (ed.) *HCCI and CAI Engines for the Automotive Industry* (Woodhead Publishing, 2007). https://www.elsevier.com/books/hcci-and-cai-engines-for-the-automotive-industry/zhao/978-1-84569-128-8
3. S. Onishi, S.H. Jo, K. Shoda, P.D. Jo, S. Kato, Active Thermo-atmospheric combustion (ATAC)—a new combustion process for internal combustion engines. SAE Tech. Paper 790501 (1979). https://doi.org/10.4271/790501
4. M. Noguchi, Y. Tanaka, T. Tanaka, Y. Takuchi, A study on gasoline engine combustion by observation of intermediate reactive products during combustion. SAE Tech. Paper 790840 (1979). https://doi.org/10.4271/790840
5. P.M. Najt, D.E. Foster, Compression ignited homogeneous charge combustion. SAE Tech. Paper 830264 (1983). https://doi.org/10.4271/830264
6. A. Oppenheim, The knock syndrome—its cures and its victims. SAE Tech. Paper 841339 (1984). https://doi.org/10.4271/841339
7. F. Zhao, D.N. Assanis, T.N. Asmus, J.E. Dec, J.A. Eng, P.M. Najt (eds.) *Homogeneous Charge Compression Ignition (HCCI) Engines: Key Research and Development Issues* (SAE International, 2003). https://www.sae.org/publications/books/content/pt-94/
8. M. Yao, Z. Zheng, H. Liu, Progress and recent trends in homogeneous charge compression ignition (HCCI) engines. Progress Energy Combust. Sci. **35**, 398–437 (2009). https://doi.org/10.1016/j.pecs.2009.05.001

9. M. Mofijur, M.M. Hasan, T.M.I. Mahlia, S.M. Ashrafur Rahman, A.S. Silitonga, H.C. Ong, Performance and emission parameters of homogeneous charge compression ignition (HCCI) engine: a review. Energies **12**, 3557 (2019). https://www.mdpi.com/1996-1073/12/18/3557

10. X. Lu, D. Han, Z. Huang, Fuel design and management for the control of an advanced compression-ignition combustion modes. Prog. Energy Combust. Sci. **37**, 741–783 (2011). https://doi.org/10.1016/j.pecs.2011.03.003

11. S. Saxena, I.D. Bedoya, Fundamental phenomena affecting low temperature combustion and HCCI engines, high load limits and strategies for extending these limits. Prog. Energy Combust. Sci. **39**, 457–488 (2013). https://doi.org/10.1016/j.pecs.2013.05.002

12. A.K. Agarwal, A.P. Singh, R.K. Maurya, Evolution, challenges and path forward for low temperature combustion engines. Prog. Energy Combust. Sci. **61**, 1–56 (2017). https://doi.org/10.1016/j.pecs.2017.02.001

13. T. Pachiannan, W. Zhong, S. Rajkumar, Z. He, X. Leng, Q. Wang, A literature review of fuel effects on performance and emission characteristics of low-temperature combustion strategies. Appl. Energy **251**, 113380 (2019). https://doi.org/10.1016/j.apenergy.2019.113380

14. X. Duan, M.-C. Lai, M. Jansons, G. Guo, J. Liua, A review of controlling strategies of the ignition timing and combustion phase in homogeneous charge compression ignition (HCCI) engine. Fuel **285**, 119142 (2021). https://doi.org/10.1016/j.fuel.2020.119142

15. M. Krishnamoorthi, R. Malayalamurthi, Z. He, S. Kandasamy, A review on low temperature combustion engines: performance, combustion and emission characteristics. Renew. Sustain. Energy Revs. **116**, 109404 (2019). https://doi.org/10.1016/j.rser.2019.109404

16. A. Mogra, K.K. Gupta, A review on new technology in internal combustion engine—HCCI engine. IOP Conf. Ser.: Mater. Sci. Eng. **810**, 012054 (2020). https://iopscience.iop.org/article/10.1088/1757-899X/810/1/012054

17. J.A. Eng, Characterization of pressure waves in HCCI combustion. SAE Tech. Paper 2002-01-2859 (2002). https://doi.org/10.4271/2002-01-2859

18. J. Dec, Y. Yang, Boosted HCCI for high power without engine knock and with ultra-low NO_x emissions—using conventional gasoline. SAE Int. J. Engines **3**(No. 2010-01-1086), 750–767 (2020). https://doi.org/10.4271/2010-01-1086

19. I.D. Bedoya, S. Saxena, F.J. Cadavid, R.W. Dibble, M. Wissink, Experimental evaluation of strategies to increase the operating range of a biogas-fueled HCCI engine for power generation. Appl. Energy **97**, 618–629 (2012). https://doi.org/10.1016/j.apenergy.2012.01.008

20. I.D. Bedoya, S. Saxena, F.J. Cadavid, R.W. Dibble, M. Wissink, Experimental study of biogas combustion in an HCCI engine for power generation with high indicated efficiency and ultra-low NO_x emissions. Energy Convers. Manage. **53**, 154–162 (2012). https://doi.org/10.1016/j.enconman.2011.08.016

21. V. Manente, B. Johansson, P. Tunestal, W. Cannella, Influence of inlet pressure, EGR, combustion phasing, speed and pilot ratio on high load gasoline partially premixed combustion. SAE Tech. Paper 2010-01-1471 (2010). https://doi.org/10.4271/2010-01-1471

22. T. Ryan, R.R. Maly, Fuel effects on engine combustion and emissions, in *Flow and Combustion in Reciprocating Engines. Experimental Fluid Mechanics*, ed. by C. Arcoumanis, T. Kamimoto (Springer, Berlin, 2008). https://doi.org/10.1007/978-3-540-68901-0_8

23. G.D. Neely, S. Sasaki, Y. Huang, J.A. Leet, D.W. Stewart, New diesel emission control strategy to meet US Tier 2 emissions regulations. SAE Tech. Paper 2005-01-1091 (2005). https://doi.org/10.4271/2005-01-1091

24. T. Kamimoto, M. Bae, High combustion temperature for the reduction of particulate in diesel engines. SAE Tech. Paper 880423 (1988). https://doi.org/10.4271/880423

25. K. Akihama, Y. Takatori, K. Inagaki, S. Sasaki, A.M. Dean, Mechanism of the smokeless rich diesel combustion by reducing temperature. SAE Tech. Paper 2001-01-0655 (2001). https://doi.org/10.4271/2001-01-0655

26. H. Yanagihara, Y. Sato, J. Mizuta, A simultaneous reduction of NO_x and soot in diesel engines under a new combustion (Uniform bulky combustion system UNIBUS), in *17th International Vienna Motor Symposium* (1996), pp. 303–314 (Nensho-Kenkyu 107, 61–74 Combust. Soc. Jpn. (1997))

27. S. Kimura, Y. Ogawa, Y. Matsui, Y. Enomoto, An experimental analysis of low temperature and premixed combustion for simultaneous reduction of NO_x and particulate emissions in direct injection diesel engines. Int. J. Engine Res. **3**, 249–259 (2002). https://doi.org/10.1243/146808702762230932

28. T. Aoyama, Y. Hattori, J. Mizuta, Y. Sato, An experimental study on premixed charge compression ignition gasoline engine. SAE Tech. Paper 960081 (1996). https://doi.org/10.4271/960081

29. K. Ohsawa, T. Kamimoto, Advanced diesel combustion, in *Flow and Combustion In Reciprocating Engines. Experimental Fluid Mechanics*, ed. by C. Arcoumanis, T. Kamimoto (Springer, Berlin, 2008). https://doi.org/10.1007/978-3-540-68901-0_7

30. Y. Takeda, K. Nakagome, K. Niimura, Emission characteristics of premixed lean diesel combustion with extremely early staged fuel injection. SAE Tech. Paper 961163 (1996). https://doi.org/10.4271/961163

31. Y. Iwabuchi, K. Kawai, T. Shoji, Y. Takeda, Trial of new concept diesel combustion system—premixed compression-ignited combustion. SAE Tech. Paper 1999-01-0185 (1999). https://doi.org/10.4271/1999-01-0185

32. T. Fang, R. Coverdill, C. Lee, R. White, Low temperature combustion within a small bore high speed direct injection (HSDI) diesel engine. SAE Tech. Paper 2005-01-0919 (2005). https://doi.org/10.4271/2005-01-0919

33. S.W. Park, R.D. Reitz, Numerical study on the low emission window of homogeneous charge compression ignition diesel combustion. Combust. Sci. Technol. **179**, 2279–2307 (2007). https://doi.org/10.1080/00102200701484142

34. S. Gan, H.K. Ng, K.M. Pang, Homogeneous charge compression ignition (HCCI) combustion: implementation and effects on pollutants in direct injection diesel engines. Appl. Energy **88**, 559–567 (2011). https://doi.org/10.1016/j.apenergy.2010.09.005

35. M.P.B. Musculus, P.C. Miles, L.M. Pickett, Conceptual models for partially premixed low-temperature diesel combustion. Prog. Energy Combust. Sci. **39**, 246–283 (2013). https://doi.org/10.1016/j.pecs.2012.09.001

36. X. Liang, Z. Zheng, H. Zhang, Y. Wang, H. Yu, A review of early injection strategy in premixed combustion engines. Appl. Sci. **9**, 3737 (2019). https://doi.org/10.3390/app9183737

37. M.L. Wyszyuski, A. Megaritis, J. Karlovsky, D. Yap, S. Peucheret, R.S. Lehrle, H. Xu, S. Golunski, Facilitation of HCCI combustion of biogas at moderate compression ratios by applications of fuel reforming and inlet air heating. J. KONES **11**, 347–356 (2004). https://www.infona.pl/resource/bwmeta1.element.baztech-article-BUJ6-0024-0077

38. S. Swami Nathan, J.M. Mallikrajuna, A. Ramesh, Homogeneous charge compression ignition versus dual fuelling for utilizing biogas in compression ignition engines. Proc. IMechE Part D: J. Autom. Eng. **223**, 413–422 (2009). https://doi.org/10.1243/09544070JAUTO970

39. S. Swami Nathan, J.M. Mallikarjuna, A. Ramesh, An experimental study of the biogas–diesel HCCI mode of engine operation. Energy Convers. Manage. **51**, 1347–1353 (2010). https://doi.org/10.1016/j.enconman.2009.09.008

40. I.D. Bedoya, S. Saxena, F.J. Cadavid, R.W. Dibble, Exploring strategies for reducing high inlet temperature requirements and allowing optimal operating conditions in a biogas fueled HCCI engine for power generation. Trans. ASME J. Eng. Gas Turbine Power **134**, 072806 (2012). https://doi.org/10.1115/1.4006075

41. I.D. Bedoya, S. Saxena, F.J. Cadavid, R.W. Dibble, Numerical analysis of biogas composition effects on combustion parameters and emissions in biogas fuelled HCCI engines for power generation. Trans. ASME J. Eng. Gas Turbines Power **135**, 071503 (2013). https://doi.org/10.1115/1.4023612

42. D. Kozarac, D. Vuilleumier, S. Saxena, R.W. Dibble, Analysis of benefits of using internal exhaust gas recirculation in biogas-fueled HCCI engines. Energy Convers. Manage. **87**, 1186–1194 (2014). https://doi.org/10.1016/j.enconman.2014.04.085

43. D. Kozarac, I. Taritas, D. Vuilleumier, S. Saxena, R.W. Dibble, Experimental and numerical analysis of the performance and exhaust gas emissions of a biogas/n-heptane fueled HCCI engine. Energy **115**, 180–193 (2016). https://doi.org/10.1016/j.energy.2016.08.055

44. K. Sudheesh, J.M. Mallikarjuna, Diethyl ether as an ignition improver for biogas homogeneous charge compression ignition (HCCI) operation—an experimental investigation. Energy **35**, 3614–3622 (2010). https://doi.org/10.1016/j.energy.2010.04.052

45. M. Feroskhan, V. Thangavel, B. Subramanian, R.K. Sankaralingam, S. Ismail, A. Chaudhary, Effects of operating parameters on the performance, emission and combustion indices of a biogas fuelled HCCI engine. Fuel **298**, 120799 (2021). https://doi.org/10.1016/j.fuel.2021.120799

46. M. Feroskhan, S. Ismail, S.H. Pancha, Study of methane enrichment in a biogas fuelled HCCI engine. Int. J. Hydrogen Energy (online, 2021). https://doi.org/10.1016/j.ijhydene.2021.02.216

47. N. Mishra, S. Mitra, A. Thapliyal, A. Mahajan, M. Feroskhan, Modelling of biogas fueled HCCI engine for various inlet conditions, in *International Conference on Emerging Trends in Engineering (ICETE)*, pp. 394–403 (Springer International Publishing, Berlin, 2020). https://www.springerprofessional.de/en/modelling-of-biogas-fueled-hcci-engine-for-various-inlet-conditi/17009284

48. W. Yoon, J. Park, Parametric study on combustion characteristics of virtual HCCI engine fueled with methane-hydrogen blends under low load conditions. Int. J. Hydrogen Energy **44**, 15511–15522 (2019). https://doi.org/10.1016/j.ijhydene.2019.04.137

49. M. Alrbai, S. Al-Dahidi, M. Abusorra, Investigation of the main exhaust emissions of HCCI engine using a newly proposed chemical reaction mechanism for biogas fuel. Case Studies Therm. Eng. **26**, 100994 (2021). https://doi.org/10.1016/j.csite.2021.100994

50. A. Mariani, A. Unich, M. Minale, EGR strategy for NO_x emission reduction in a CAI engine fuelled with innovative biogas. Tec. Italian J. Eng. Sci. **63**, 417–423 (2019). https://doi.org/10.18280/ti-ijes.632-444

51. A. Mariani, M. Minale, A. Unich, Use of biogas containing CH_4, H_2 and CO_2 in controlled auto-ignition engines to reduce NO_x emissions. Fuel **301**, 120925 (2021). https://doi.org/10.1016/j.fuel.2021.120925

52. E. Tomita, N. Kawahara, Z. Piao, R. Yamaguchi, Effects of EGR and early injection of diesel fuel on combustion characteristics and exhaust emissions in a methane dual fuel engine. SAE Tech. Paper 2002-01-2723 (2002). https://doi.org/10.4271/2002-01-2723

53. S. Singh, S.R. Krishnan, K.K. Srinivasan, K.C. Midkiff, S.R. Bell, Effect of pilot injection timing, pilot quantity and intake charge conditions on performance and emissions for an advanced low-pilot-ignited natural gas engine. Int. J. Engine Res. **5**, 329–348 (2004). https://doi.org/10.1243/146808704323224231

54. T. Ishiyama, H. Kawanabe, K. Ohashi, M. Shioji, S. Nakai, A study on premixed charge compression ignition combustion of natural gas with direct injection. Int. J. Engine Res. **6**, 443–451 (2005). https://doi.org/10.1243/146808705X30459

55. K. Inagaki, T. Fuyuto, K. Nishikawa, K. Nakakita, I. Sakata, Dual-fuel PCI combustion controlled by in-cylinder stratification of ignitability. SAE Tech. Paper 2006-01-0028 (2006). https://doi.org/10.4271/2006-01-0028

56. S.L. Kokjohn, R.M. Hanson, D.A. Splitter, R.D. Reitz, Fuel reactivity controlled compression ignition (RCCI): a pathway to controlled high-efficiency clean combustion. Int. J. Engine Res. **12**, 209–226 (2011). https://doi.org/10.1177/1468087411401548

57. R.D. Reitz, G. Duraisamy, Review of high efficiency and clean reactivity controlled compression ignition (RCCI) combustion in internal combustion engines. Prog. Energy Combust. Sci. **46**, 12–71 (2015). https://doi.org/10.1016/j.pecs.2014.05.003

58. J. Li, W. Yang, D. Zhou, Review on the management of RCCI engines. Renew. Sustain. Energy Rev. **69**, 65–79 (2017). https://doi.org/10.1016/j.rser.2016.11.159

59. I.B. Dalha, M.A. Said, Z.A. Abdul Karim, F. Firmansyah, Strategies and methods of RCCI combustion: a review. AIP Conf. Proc. **2035**, 030006 (2018). https://doi.org/10.1063/1.5075562

60. D. Splitter, M. Wissink, D. DelVescovo, R. Reitz, RCCI engine operation towards 60% thermal efficiency. SAE Tech. Paper 2013-01-0279 (2013). https://doi.org/10.4271/2013-01-0279

61. E. Shim, H. Park, C. Bae, Comparisons of advanced combustion technologies (HCCI, PCCI, and dual-fuel PCCI) on engine performance and emission characteristics in a heavy-duty diesel engine. Fuel **262**, 116436 (2020). https://doi.org/10.1016/j.fuel.2019.116436

62. M.M. Ibrahim, J.V. Narasimhan, A. Ramesh, Comparison of the predominantly premixed charge compression ignition and the dual fuel modes of operation with biogas and diesel as fuels. Energy **89**, 990–1000 (2015). https://doi.org/10.1016/j.energy.2015.06.033

63. S.H. Park, S.M. Yoon, Effect of dual-fuel combustion strategies on combustion and emission characteristics in reactivity controlled compression ignition (RCCI) engine. Fuel **181**, 310–318 (2016). https://doi.org/10.1016/j.fuel.2016.04.118

64. M. Ebrahimi, S.A. Jazayer, Effect of hydrogen addition on RCCI combustion of a heavy duty diesel engine fueled with landfill gas and diesel oil. Int. J. Hydrogen Energy **44**, 7607–7615 (2019). https://doi.org/10.1016/j.ijhydene.2019.02.010

65. K.A. Rahman, A. Ramesh, Effect of reducing the methane concentration on the combustion and performance of a biogas diesel predominantly premixed charge compression ignition engine. Fuel **206**, 117–132 (2017). https://doi.org/10.1016/j.fuel.2017.05.100

66. I.B. Dalha, M.A. Said, Z.A.A. Karim, M. El-Adawy, Effects of port mixing and high carbon dioxide contents on power generation and emission characteristics of biogas-diesel RCCI combustion. Appl. Therm. Eng. **198**, 117449 (2021). https://doi.org/10.1016/j.appltherm aleng.2021.117449

67. P. Asadollahzadeh, M.H. Hamedi, S.A. Jazayeri, A comprehensive study of a reactivity-controlled compression ignition engine fueled with biogas and diesel oil. Clean Technol. Environ. Policy **23**, 113–126 (2021). https://doi.org/10.1007/s10098-020-01983-z

68. D. Robertson, R. Prucka, A review of spark-assisted compression ignition (SACI) research in the context of realizing production control strategies. SAE Tech. Paper 2019-24-0027 (2019). https://doi.org/10.4271/2019-24-0027

69. J. Hyvönen, G. Haraldsson, B. Johansson, Operating conditions using spark assisted HCCI combustion during combustion mode transfer to SI in a multi-cylinder VCR-HCCI engine. SAE Tech. Paper 2005-01-0109 (2005). https://doi.org/10.4271/2005-01-0109

70. T. Urushihara, K. Yamaguchi, K. Yoshizawa, T. Itoh, A study of a gasoline-fueled compression ignition engine—expansion of HCCI operation range using SI combustion as a trigger of compression ignition. SAE Tech. Paper 2005-01-0180 (2005). https://doi.org/10.4271/2005-01-0180

71. M. Kawano, T. Urushihara, M. Sueoka, A. Inoue, M. Nishida, Y. Nakahara, M. Koutoku, H. Yokohata, Combustion technology of new generation gasoline engine with spark controlled compression ignition. J. Combust. Soc. Jpn. **62**, 204–211 (2020). https://doi.org/10.20619/jco mbsj.62.201_204

72. U. Azimov, E. Tomita, N. Kawahara, Y. Harada, Premixed mixture ignition in the end-gas region (PREMIER) combustion in a natural gas dual-fuel engine: operating range and exhaust emissions. Int. J. Engine Res. **12**, 484–497 (2011). http://jer.sagepub.com/content/12/5/484

73. C. Aksu, N. Kawahara, K. Tsuboi, S. Nanba, E. Tomita, M. Kondo, Effect of hydrogen concentration on engine performance, exhaust emissions and operation range of PREMIER combustion in a dual fuel gas engine using methane-hydrogen mixtures. SAE Tech. Paper 2015-01-1792 (2015). https://doi.org/10.4271/2015-01-1792

74. U. Azimov, E. Tomita, N. Kawahara, Ignition, Combustion and exhaust emission characteristics of micro-pilot ignited dual-fuel engine operated under PREMIER combustion mode. SAE Tech. Paper 2011-01-1764 (2011). https://doi.org/10.4271/2011-01-1764

75. C. Aksu, N. Kawahara, K. Tsuboi, M. Kondo, E. Tomita, Extension of PREMIER combustion operation range using split micro pilot fuel injection in a dual fuel natural gas compression ignition engine: a performance-based and visual investigation. Fuel **185**, 243–253 (2016). https://doi.org/10.1016/j.fuel.2016.07.120

76. A. Valipour Berenjestanaki, N. Kawahara, K. Tsuboi, E. Tomita, End-gas autoignition characteristics of PREMIER combustion in a pilot fuel-ignited dual-fuel biogas engine. Fuel **254**, 115634 (2019). https://doi.org/10.1016/j.fuel.2019.115634

77. A. Valipour Berenjestanaki, N. Kawahara, K. Tsuboi, E. Tomita, Performance, emissions and end-gas autoignition characteristics of PREMIER combustion in a pilot fuel-ignited dual-fuel biogas engine with various CO_2 ratios. Fuel **286**, 119330 (2021). https://doi.org/10.1016/j.fuel.2020.119330

78. U. Azimov, E. Tomita, N. Kawahara, Y. Harada, Effect of syngas composition on combustion and exhaust emission characteristics in a pilot-ignited dual-fuel engine operated in PREMIER combustion mode. Int. J. Hydrogen Energy **36**, 11985–11996 (2011). https://doi.org/10.1016/j.ijhydene.2011.04.192

79. E. Tomita, N. Kawahara, J. Zheng, Visualization of Auto-Ignition of End Gas Region without Knock in a Spark-Ignition Natural Gas Engine. J. KONES Powertrain Transport **17**, 521–527 (2010). http://yadda.icm.edu.pl/yadda/element/bwmeta1.element.baztech-article-BUJ7-0018-0008

80. N. Kawahara, Y. Kim, H. Wadahama, K. Tsuboi, E. Tomita, Differences between PREMIER combustion in a natural gas spark-ignition engine and knocking with pressure oscillations. Proc. Combust. Inst. **37**, 4983–4991 (2019). https://doi.org/10.1016/j.proci.2018.08.055

81. J.D. Dale, M.D. Checkel, P.R. Smy, Application of high energy ignition systems to engines. Prog. Energy Combust. Sci. **23**, 379–398 (1997). https://doi.org/10.1016/S0360-1285(97)00011-7

82. A.Y. Starikovskii, Plasma supported combustion. Proc. Combust. Inst. **30**, 2405–2417 (2005). https://doi.org/10.1016/j.proci.2004.08.272

83. A. Starikovskiy, N. Aleksandrow, Plasma-assisted ignition and combustion. Prog. Energy Combust. Sci. **39**, 61–110 (2013). https://doi.org/10.1016/j.pecs.2012.05.003

84. S.M. Starikovskaia, Plasma-assisted ignition and combustion: nanosecond discharges and development of kinetic mechanisms. J. Phys. D: Appl. Phys. **47**, 353001 (2014). https://doi.org/10.1088/0022-3727/47/35/353001

85. Y. Ju, W. Sun, Plasma assisted combustion Dynamics and chemistry. Prog. Energy Combust. Sci. **48**, 21–83 (2015). https://doi.org/10.1016/j.pecs.2014.12.002

86. Y. Ju, J.K. Lefkowitz, C.B. Reuter, S.H. Won, X. Yang, S. Yang, W. Sun, Z. Jiang, Q. Chen, Plasma assisted low temperature combustion. Plasma Chem. Plasma Process. **36**, 85–105 (2016). https://doi.org/10.1007/s11090-015-9657-2

87. R. Hickling, W.R. Smith, Combustion bomb tests of laser ignition. SAE Tech. Paper 740114 (1974). https://doi.org/10.4271/740114

88. J.D. Dale, P.R. Smy, *The First Laser Ignition Engine Experiment (c.a. 1976) in Laser Ignition Conference, OSA Tech. Digest* (online) (Optical Society of America, 2015), paper T3A.1 (1976). https://doi.org/10.1364/LIC.2015.T3A.1

89. J.D. Dale, P.R. Smy, R.M. Clements, Laser ignited internal combustion engine—an experimental study. SAE Tech. Paper 780329 (1978). https://doi.org/10.4271/780329

90. P.D. Ronney, Laser versus conventional ignition of flames. Opt. Eng. **33**, 510–521 (1994). https://doi.org/10.1117/12.152237

91. D. Bradley, C.G.W. Sheppard, I.M. Suardjaja, R. Woolley, Fundamentals of high-energy spark ignition with lasers. Combust. Flame **138**, 55–77 (2004). https://doi.org/10.1016/j.combustflame.2004.04.002

92. T.X. Phuoc, Single-point versus multi-point laser ignition: experimental measurements of combustion times and pressures. Combust. Flame **122**, 508–510 (2000). https://doi.org/10.1016/S0010-2180(00)00137-1

93. T.X. Phuoc, Laser-induced spark ignition fundamental and applications. Opt. Lasers Eng. **44**, 351–397 (2006). https://doi.org/10.1016/j.optlaseng.2005.03.008

94. M.H. Morsy, Review and recent developments of laser ignition for internal combustion engines applications. Renew. Sustain. Energy Rev. **16**, 4849–4875 (2012). https://doi.org/10.1016/j.rser.2012.04.038

95. T. Nakayama, N. Kawahara, E. Tomita, Y. Ikeda, High temporally resolved optical measurement for laser ignition process of laminar premixed mixture. Trans. Jpn. Soc. Mech. Engr. B **74**, 1633–1640 (2008). https://doi.org/10.1299/kikaib.74.1633

96. J.L. Beduneau, N. Kawahara, T. Nakayama, E. Tomita, Y. Ikeda, Laser-induced radical generation and evolution to a self-sustaining flame. Combust. Flame **156**, 642–656 (2009). https://doi.org/10.1016/j.combustflame.2008.09.013

97. J.D. Mullett, P.B. Dickinson, A.T. Shenton, G. Dearden, K.G. Watkins, Multi-cylinder laser and spark ignition in an IC gasoline automotive engine: a comparative study. SAE Tech. Paper 2008-01-0470 (2008). https://doi.org/10.4271/2008-01-0470

98. M. Weinrotter, H. Kopecek, E. Wintner, Laser ignition of engines. Laser Phys. **15**, 947–953 (2005)

99. H. Kofler, J. Tauer, G. Tartar, K. Iskra, J. Klausner, G. Herdin, E. Wintner, An innovative solid-state laser for engine ignition. Laser Phys. Lett. **4**, 322–327 (2007). https://doi.org/10.1002/lapl.200610106

100. M. Tsunekane, T. Inohara, A. Ando, N. Kido, K. Kanehara, T. Taira, High peak power, passively Q-switched microlaser for ignition of engines. IEEE J Quantum Electron. **46**, 277–284 (2010). https://ieeexplore.ieee.org/document/5401116

101. M. Tsunekane, T. Inohara, K. Kanehara, T. Taira, Chapter 10: micro-solid-state laser for ignition of automobile engines, in *Advances in Solid-State Lasers: Development and Applications*, ed. pp. 195–212, ed. by M. Grishin (InTech Web Publisher, 2010). https://www.intechopen.com/chapters/8409

102. G. Kroupa, G. Franz, E. Winkelhofer, Novel miniaturized high-energy Nd-YAG laser for spark ignition in internal combustion engines. Opt. Eng. **48**, 014202 (2009). https://doi.org/10.1117/1.3072958

103. E. Schwarz, I. Muri, J. Tauer, H. Kofler, E. Wintner, Laser-induced ignition by optical breakdown. Laser Phys. **20**, 1545–1553 (2010). https://doi.org/10.1134/S1054660X10110204

104. J. Tauer, H. Kofler, E. Wintner, Laser-initiated ignition. Laser Photon Rev. **4**, 99–122 (2010). https://doi.org/10.1002/lpor.200810070

105. T. Saito, K. Yanagisawa, H. Furutani, Gasoline engine performance with laser-induced breakdown ignition under EGR condition, in *1st Laser Ignition Conference* (Yokohama, Japan) LIC9-4 (2013)

106. Y. Ma, X. Li, X. Yu, R. Fan, R. Yan, J. Peng, X. Xu, R. Sun, D. Chen, A novel miniaturized passively Q-switched pulse-burst laser for engine ignition. Opt. Express **22**, 24655–24665 (2014). https://doi.org/10.1364/OE.22.024655

107. N. Pavel, M. Tsunekane, T. Taira, Composite, all-ceramics, high-peak-power Nd:YAG/Cr4⁺: YAG monolithic micro-laser with multiple-beam output for engine ignition. Opt. Express **19**, 9378–9384 (2011). https://doi.org/10.1364/OE.19.009378

108. N. Pavel, M. Bärwinkel, P. Heinz, D. Brüggemann, G. Dearden, G. Croitoru, O.V. Grigore, Laser ignition—spark plug development and application in reciprocating engines. Prog. Quantum Electron. **58**, 1–32 (2018). https://doi.org/10.1016/j.pquantelec.2018.04.001

109. N. Pavel, R. Chiriac, A. Birtas, F. Draghici, M. Dinca, On the improvement by laser ignition of the performances of a passenger car gasoline engine. Opt. Express **27**, A385–A396 (2019). https://doi.org/10.1364/OE.27.00A385

110. S.B. Gupta, M. Biruduganti, B. Bihari, R. Sekar, Chapter 10: natural gas fired reciprocating engines for power generation: concerns and recent advances, in *Natural Gas—Extraction to End Use*, pp. 211–234, ed. by S.B. Gupta. (InTech web publisher, 2012) https://www.intechopen.com/chapters/40563

111. B. Almansour, S. Vasu, S.B. Gupta, Q. Wang, R. Van Leeuwen, C. Ghosh, Performance of a laser ignited multicylinder lean burn natural gas engine. Trans. ASME J. Eng. Gas Turbines Power **139**, 111501 (2017). https://doi.org/10.1115/1.4036621

112. M.A.V. Ward, T.T. Wu, A theoretical study of the microwave heating of a cylindrical shell, flame-front electron plasma in an internal combustion engine. Combust. Flame **32**, 57–71 (1978). https://doi.org/10.1016/0010-2180(78)90080-9

113. E.G. Groff, M.K. Krage, Microwave effects on premixed flames. Combust. Flame **56**, 293–306 (1984). https://doi.org/10.1016/0010-2180(84)90063-4

114. X. Rao, K. Hemawan, I. Wichman, C. Carter, T. Grotjohn, J. Asmussen, T. Lee, Combustion dynamics for energetically enhanced flames using direct microwave energy coupling. Proc. Combust. Inst. **33**, 3233–3240 (2011). https://doi.org/10.1016/j.proci.2010.06.024

115. W. Sun, M. Uddi, S. Won, T. Ombrello, C. Carter, Y. Ju, Kinetic effects of non-equilibrium plasma-assisted methane oxidation on diffusion flame extinction limits. Combust. Flame **159**, 221–229 (2012). https://doi.org/10.1016/j.combustflame.2011.07.008

116. Y. Ikeda, A. Nishiyama, Y. Wachi, M. Kaneko, Research and development of microwave plasma combustion engine, in *Part I: Concept of Plasma Combustion and Plasma Generation Technique*. SAE Tech. Paper 2009-01-1050 (2009). https://doi.org/10.4271/2009-01-1050

117. Y. Ikeda, A. Nishiyama, H. Katano, M. Kaneko, H. Jeong, Research and development of microwave plasma combustion engine, in *Part II: Engine Performance of Plasma Combustion Engine*. SAE Tech. Paper 2009-01-1049 (2009). https://doi.org/10.4271/2009-01-1049

118. A. Nishiyama, Y. Ikeda, Improvement of lean limit and fuel consumption using microwave plasma ignition technology. SAE Tech. Paper 2012-01-1139 (2012). https://doi.org/10.4271/2012-01-1139

119. A. DeFilippo, S. Saxena, V. Rapp, R. Dibble, J.Y. Chen, A. Nishiyama, Y. Ikeda, Extending the lean stability limits of gasoline using a microwave-assisted spark plug. SAE Tech. Paper 2011-01-0663 (2011). https://doi.org/10.4271/2011-01-0663

120. J. Hwang, W. Kim, C. Bae, Influence of plasma-assisted ignition on flame propagation and performance in a spark-ignition engine. Appl. Energy Combust. Sci. **6**, 100029 (2021). https://doi.org/10.1016/j.jaecs.2021.100029

121. B. Wolk, A. DeFilippo, J.-Y. Chen, R. Dibble, A. Nishiyama, Y. Ikeda, Enhancement of flame development by microwave-assisted spark ignition in constant volume combustion chamber. Combust. Flame **160**, 1225–1234 (2013). https://doi.org/10.1016/j.combustflame.2013.02.004

122. Y. Ikeda, S. Padala, M. Makita, A. Nishiyama, Development of innovative microwave plasma ignition system with compact microwave discharge igniter. SAE Tech. Paper 2015-24-2434 (2015). https://doi.org/10.4271/2015-24-2434

123. M.K. Le, A. Nishiyama, T. Serizawa, Y. Ikeda, Applications of a multi-point microwave discharge igniter in a multi-cylinder gasoline engine. Proc. Combust. Inst. **37**, 5621–5628 (2019). https://doi.org/10.1016/j.proci.2018.06.033

124. C.D. Cathey, T. Tang, T. Shiraishi, T. Urushihara, A. Kuthi, M.A. Gundersen, Nanosecond plasma ignition for improved performance of an internal combustion engine. IEEE Trans. Plasma Sci. **35**, 1664–1668 (2007). https://ieeexplore.ieee.org/document/4392536

125. T. Shiraishi, A. Kakuho, T. Urushihara, C. Cathey, T. Tang, M. Gundersen, A study of volumetric ignition using high-speed plasma for improving lean combustion performance in internal combustion engine. SAE Tech. Paper 2008-01-0466 (2008). https://doi.org/10.4271/2008-01-0466

126. T. Shiraishi, T. Urushihara, M.A. Gundersen, A trial of ignition innovation of gasoline engine by nanosecond pulsed low temperature plasma ignition. J. Phys, D: Appl. Phys. **42**, 135208 (2009). https://doi.org/10.1088/0022-3727/42/13/135208

127. T. Shiraishi, T. Urushihara, Fundamental analysis of combustion initiation characteristics of low temperature plasma ignition for internal combustion gasoline engine. SAE Tech. Paper 2011-01-0660 (2011). https://doi.org/10.4271/2011-01-0660

128. K. Tanoue, E. Hotta, Y. Moriyoshi, Development of a novel ignition system using repetitive pulse discharges: ignition characteristics of premixed hydrocarbon-air mixtures. SAE Tech. Paper 2008-01-0468 (2008). https://doi.org/10.4271/2008-01-0468

129. K. Tanoue, T. Kuboyama, Y. Moriyoshi, E. Hotta, Y. Imanishi, N. Shimizu, K. Iida, Development of a novel ignition system using repetitive pulse discharges: application to a SI engine. SAE Tech. Paper 2009-01-0505 (2009). https://doi.org/10.4271/2009-01-0505

130. K. Tanoue, Y. Moriyoshi, E. Hotta, Enhancement of ignition characteristics of lean premixed hydrocarbon-air mixtures by repetitive pulse discharges. Int. J. Engine Res. **10**, 399–407 (2009). https://doi.org/10.1243/14680874JER04309

131. W.F. Piock, P. Weyand, E. Wolf, V. Heise, Ignition systems for spray-guided stratified combustion. SAE Tech. Paper 2010-01-0598 (2010). https://doi.org/10.4271/2010-01-0598

132. V. Heise, P. Farah, H. Husted, E. Wolf, High frequency ignition system for gasoline direct injection engines. SAE Tech. Paper 2011-01-1223 (2011). https://doi.org/10.4271/2011-01-1223

133. T. Shiraishi, A study of low temperature plasma-assisted gasoline HCCI combustion. SAE Int. J. Engines **12**, 101–113 (2019). https://doi.org/10.4271/03-12-01-0008
134. A. Mariani, F. Foucher, Radio frequency spark plug: an ignition system for modern internal combustion engines. Appl. Energy **122**, 151–161 (2014). https://doi.org/10.1016/j.apenergy.2014.02.009
135. J.B. Heywood, *Internal combustion engine fundamentals* (2nd ed.) (McGraw Hill, 2018). https://www.accessengineeringlibrary.com/content/book/9781260116106
136. R. Miller, *Supercharged Engine*. United States Patent Office 2,817,322 (1957)
137. Cross-ministerial Strategic Innovation Promotion (SIP) Program, *Innovative Combustion Technology (2019)*. https://www.jst.go.jp/sip/dl/k01/k01_seika2019.pdf (in Japanese) Accessed 19 Nov 2021

Printed in the United States
by Baker & Taylor Publisher Services